读客文化

| 中英双语版 |

万物简介
星系是什么

[英] 约翰·格里布本 著

孙红贵 译

GALAXIES:
A VERY SHORT INTRODUCTION

浙江科学技术出版社

著作合同登记号 图字：11-2022-39

Galaxies: A Very Short Introduction
Copyright: © John and Mary Gribbin 2008
Galaxies: A Very Short Introduction was originally published in English in 2008. This translation is published by arrangement with Oxford University Press. Dook Media Group Limited is solely responsible for this translation from the original work and Oxford University Press shall have no liability for any errors, omissions or inaccuracies or ambiguities in such translation or for any losses caused by reliance thereon

图书在版编目（CIP）数据

万物简介．星系是什么：汉文、英文 /（英）约翰·格里布本 (John Gribbin) 著；孙红贵译．— 杭州：浙江科学技术出版社，2022.11

书名原文：Galaxies：A Very Short Introduction
ISBN 978-7-5341-9971-4

Ⅰ．①万… Ⅱ．①约… ②孙… Ⅲ．①星系－普及读物－汉、英 Ⅳ．① P15-49

中国版本图书馆 CIP 数据核字 (2022) 第 051162 号

书　　名　万物简介：星系是什么
著　　者　［英］约翰·格里布本
译　　者　孙红贵

出　　版　浙江科学技术出版社
网　　址　www.zkpress.com
地　　址　杭州市体育场路 347 号
联系电话　0571-85176593
邮政编码　310006
印　　刷　三河市龙大印装有限公司
发　　行　读客文化股份有限公司

开　　本　880mm × 1230mm 1/32
印　　张　9.5
字　　数　152 000
版　　次　2022 年 11 月第 1 版
印　　次　2022 年 11 月第 1 次印刷
书　　号　ISBN 978-7-5341-9971-4
定　　价　49.90 元

责任编辑　卢晓梅　　**责任校对**　张　宁
责任美编　金　晖　　**责任印务**　叶文炀

推荐序

在晴朗且没有光污染的夜晚，我们抬头望向天空，便会看见群星闪烁。自古以来，这片星空就使人们充满好奇：这些星星是什么，离我们有多远？星星之外又是什么？宇宙有多大？一代又一代的人为探索这些问题而努力着。到今天，我们已经知道，夜晚天空中肉眼可见的绝大部分星星都是像太阳一样的恒星，它们和我们人类都处在一个叫银河系的星系内。同时，银河系也只是一个普通的星系，和上亿个星系一起，构成了我们的宇宙。

《万物简介：星系是什么》是由牛津大学出版的通识读物，作者是约翰·格里布本博士，他毕业于剑桥大学。该书从银河系内恒星距离测量开始，比较完整地介绍了银河系的空间尺度、最近的河外星系——

仙女座星系的发现、宇宙的膨胀、星系的起源和分类、宇宙未来的命运和星系的演化等内容。作者用生动、简洁的文字对星系和宇宙演化的相关成果进行了介绍，内容较全面，是一本非常难得的、具备专业水平的同时，又有很强科普性的读物，完全适合高中以上水平的读者阅读。本书的中文翻译也非常准确，完整保留了英文的内容和风格。作为本书的审校者，我对原著作中出现的少数表述不清的地方进行了说明，并更新了部分观测数据。本书总体的修改非常少，力求最大程度保留原著的风格。

宇宙与人类的关系，是我们永恒的探索话题。当下，生活和工作节奏都非常快，又笼罩着疫情的阴影，更需要我们深入思考人类和自然界的关系。我郑重推荐《万物简介：星系是什么》这本科普书，希望它给您带来一份心灵的安静，和对宇宙虔诚的热爱。

2022.07.14 于南京仙林湖畔

康熙

引言

英文页码 1

人类真正开始对星系进行科学研究是在20世纪20年代，那时，人们首次意识到，透过望远镜看到的许多模糊光斑，其实是太空中大量恒星组成的、远在银河系之外的宇宙岛。如果没有望远镜，人类永远不可能探索银河系以外的宇宙，当然也不可能洞悉河外星系的基本特征。事实上，望远镜是在历经了近400年的技术发展之后，才精良到足以让人一点点看清星系的真面目。

据说，第一个使用望远镜观察夜空的人是伦纳德·迪格斯（Leonard Digges）。他是一位数学家、勘测员，毕业于牛津大学。1551年前后，伦纳德发明了经纬仪。考虑到经纬仪对他的工作太重要，对望远镜（本质上就是朝向天空的经纬仪）的这种用法他一直守口如瓶。但伦纳德写了一本书，该书算是第一批科普英文读物之一，书中包含了托勒密地心说的宇宙模型。伦纳德死于1559年，他的儿子托马斯·迪格斯（Thomas Digges）继承了父亲的事业。托马斯出生于16世纪40年代，后来成为一名数学家，并于1571年整理出版了父亲的一本遗著，书中首次披露了望远镜这一秘密。托马斯也从事天文观测，并于1576年出版了他父亲第一本著作的修订扩充版，书中首次用英文介绍了哥白尼日心说的宇宙模型。

2

在那本名为《永恒的预言》（*Prognostication Everlasting*）的书中，托马斯声称宇宙是无限的，并附了一幅插图。图中，被行星环绕的太阳端坐在向四面八方无限延伸的恒星阵列的中心。我们知道托马斯至少有一架望远镜，因此可以合理地推断，他完全有可能用望远镜观察到了天空中那条神秘的光带，并发现银河系是由数不清的恒星组成的。

迪格斯父子的故事可能会让人感到有些不可思议，因为人们通常认为第一个使用天文望远镜的是伽利略·伽利雷（Galileo Galilei），并确信17世纪头十年，伽利略率先在望远镜中观察到银河系是由大量恒星组成的。事实上，望远镜在欧洲西北部被不同的人独立发明了好几次，直到1609年才从荷兰传到意大利。据说，伽利略制造了好几架望远镜。不过，带有传奇色彩的是，伽利略仅凭别人对望远镜工作原理的描述，就成功制作了属于他的第一架望远镜，并且马上就将镜筒对准了天空。伽利略将自己的发现记录在1610年出版的《星际使者》(*Sidereus Nuncius*，英文名为 *The Starry Messenger*）一书中。这使伽利略声名鹊起，他也因此被误认为是第一个使用望远镜的天文学家。像之前的托马斯·迪格斯一样，伽利略无疑也确实观察到了银河系是由无数恒星组成的。

下一个有关我们在宇宙中所处位置的成就，是由英国仪器制造商、哲学家托马斯·赖特（Thomas Wright）在18世纪中叶取得的。但是，就像迪格斯父子一样，他的贡献也基本上被人遗忘了。众所周知，银河系看上去就像一条横贯夜空的光带。然而，在1750年出版的《宇宙的原始理论或新假说》(*An Original Theory or New Hypothesis of the Universe*）一书中，赖特指出，银河系是由一片平铺的恒星组成的，并将其形状比作磨坊里的磨盘。更令人印象深刻的是，赖特意识到太阳并不位于这个圆盘的中心，而是偏向一边。尽管赖特未能实现想象的飞跃，进一步说明星云可能就是与银河系相似的恒星系统，但赖特的确提到，通过望远镜看到的模糊光斑，也就是云状的星云，可能位于银河系之外。另一位哲学家伊曼努尔·康德（Immanuel Kant）基于赖特的观点，又向前迈出了一大步。康德指出，星云很可能是像银河系一样的“宇宙岛”。但这个想法在当时并没有得到人们足够的重视。

随着望远镜的改进，新发现并编入目录的星云越来越多。精心编制星云目录的原因之一是彗星热。18 世纪末和 19 世纪初，天文学家曾热衷于搜寻彗星。然而星云和彗星看上去都是模糊的光斑，不仔细辨别很容易混淆。因此有些人，比如 18 世纪 80 年代的查尔斯·梅西耶（Charles Messier）和 1802 年完成了一个目录编制的威廉·赫歇尔（William Herschel），都试图通过编制目录标定星云的位置，以免发生混淆。赫歇尔的目录涵盖了 2500 个星云，现在已知其中大部分是星系。在接下来的 20 年里，赫歇尔曾试图弄清楚星云是由什么构成的。但即使动用当时最大的、口径达 48 英寸（约 1.2 米）的望远镜，也无法将这些模糊的光斑分解成恒星。直到 1822 年去世时，赫歇尔仍坚信星云是银河系中弥散的物质云。

下一个观测成就是由第三代罗斯伯爵威廉·帕森斯（William Parsons）取得的。19 世纪 40 年代，帕森斯建造了一架口径为 72 英寸（约 1.8 米）的巨型望远镜，并借此发现许多星云都具有旋涡结构，看起来就像是一杯黑咖啡里搅出来的奶油图案。在接下来的几十年里，一些星云被确定为银河系内发光的气体云，另一些星云则被认证为星团。尽管星团的规模比银河系小得多，但明显与银河系存在关联。旋涡星云则与这两种类型天体的特征都不相符。19 世纪下半叶，天文摄影技术的发展使人们能更便捷地研究旋涡星云，但当时拍摄的照片并不足以揭示旋涡星云的本质。

4

20 世纪初，绝大多数天文学家都认为，旋涡星云是恒星形成过程中围绕恒星旋转的物质云，就像形成太阳系的物质云一样。但此后的 20 年间，支持宇宙岛观点的人越来越多，这促使美国国家科学院（National Academy of Sciences）就这一议题主办了一场辩论会。当时，来自美国加利福尼亚州威尔逊山天文台的哈罗·沙普利（Harlow Shapley）代表反对宇宙岛观点的多数派；美国加利福尼亚州利克天文台（Lick Observatory）的赫伯·柯蒂斯（Heber Curtis）则代表认同该观点的少数派。这场于 1920 年 4 月 26 日举行的辩论会被天文学家称为“大辩论”（The Great Debate）。虽然没能决出胜负，但这场大辩论标志着关于星系的现代科学研究的开端。

目录

01

大辩论

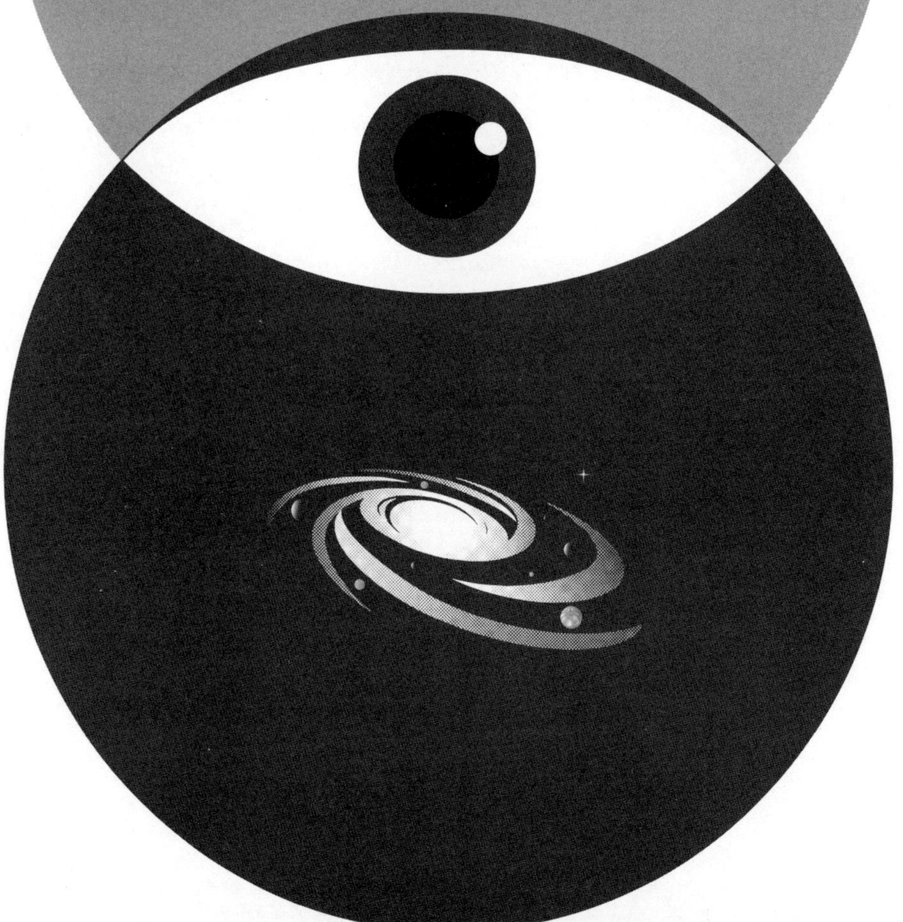

在1920年4月26日举行的天文学“大辩论”[5]中，有两个争论的焦点：一是银河系到底有多大；二是旋涡星云的本质是什么。事实上，会议中根本就没有发生真正意义上的辩论。大会上对立的双方只是分别做了40分钟的演讲，接下来是很平常的讨论。会议在史密森尼国家自然历史博物馆（当时为美国国家博物馆）举行，主题是“宇宙的尺度”。尽管沙普利和柯蒂斯在宇宙尺度的问题上存在严重分歧，但双方捍卫自己观点的方式，也仅限于在次年的学术期刊上发表各自的论文而已。本质上，沙普利认为银河系基本就是宇宙的全部，或者至少是宇宙中最重要的部分，他感兴趣的是我们银河系的尺度；柯蒂斯则认为，旋涡星云是与银河系一样独立的星系，他更关心银河系以外的空间有多么辽阔。

“大辩论”的发生正当其时，因为天文界刚研究出了一些适用于测量银河系尺度的测距技术。可以用伦纳德·迪格斯的方法——也就是三角视差法测量地球到附近恒星的距离。对那些距离比较近的恒星，经过长时间观测可以发现，它们在星空背景上有轻微的偏移。当地球在其绕太阳轨道的相对两侧，也就是位于间隔 6 个月的两个位置观察时，这种偏移达到最大。其实，只要将恒星想象成一根手指，就能解释这种视差现象。将手指放在眼前，保持不动，轮流用一只眼睛看，会发现手指相对于背景有偏移。手指离眼睛越近，视差效应就越大。与此类似，三角视差法只需知道恒星（角）位移的大小和地球轨道的直径（已从太阳系的三角测量中得知），就可以计算出恒星到地球的距离。

6

不幸的是，大多数恒星离我们太远，这种视差效应太小，以至于根本无法测量。即使是最近的恒星半人马座阿尔法星（Alpha Centauri），距离太阳也非常遥远，远到光需要 4.24 年[1] 的时间（也就是 4.24 光年远）才能穿越两者间的广袤空间。到 1908 年，用三角视差法成功测量出距离的恒星，大约只有 100 颗。另一种涉及几何的测距方法是利用离我们较近的移动星团，并假设其中的恒星一起在太空中移动，这样测量的距离可以远达 100 光年，或者，用天文学家常用的单位表示，大约为 30 秒差距（1 秒差距差不多是 3.26 光年）。这种方法使天文学家足以校准天文学中最重

---

1 原书出版于 2008 年，本书内文中数据已根据最新数据更新过。本书注释若无特别说明，均为编者注。

要的示距天体。[1]

想要明确这个新示距天体在透视法中的重要性，只要看看在20世纪早期对银河系大小做出的最佳估计就明白了。荷兰天文学家雅各布斯·卡普廷（Jacobus Kapteyn）根据前文解释过的测距法与观测到的恒星的视亮度，统计了不同方向上相同大小的天空区域内可见的恒星数量，并估算了地球与这些恒星之间的距离。他由此推断，银河系就像个铁饼，中间厚度大约有2千秒差距，直径为10千秒差距，太阳居于其中心附近。现在看来这个估算值太小了，主要是因为恒星间有大量的尘埃，但卡普廷对此并不知情。尘埃就像雾一样，使我们沿银河系平面方向看不了多远，这种现象被称为恒星消光。就像迷失在雾中的旅行者自认为站在世界中心一样，卡普廷似乎也迷失在银河系的迷雾中，封闭在自己的小宇宙中心。就在不到100年之前，大多数天文学家还依然认为，银河系就是宇宙的全部。

在20世纪的第二个十年，情况开始发生变化。哈

1 这里作者是指移动星团法，一种用运动学的手段测量距离的方法，也就是根据星团中各恒星的运动速度确定距离。应该特别指出的是，这种方法必须假定移动星团中所有的恒星都以相同速度、相同方向移动。对银河系之外的天体，即使用移动星团法也不能测定它们与地球之间的距离。

佛大学天文台的亨丽埃塔·斯旺·莱维特（Henrietta Swan Leavitt）发现一类被称为**造父变星**（Cepheid）的恒星有特殊的亮度变化方式，这使得它们可以被用作示距天体。每一颗造父变星都有规律地变亮或变暗，并精确地重复这个光变周期。有些造父变星在不到1天的时间内就完成一个光变周期，另一些则需要长达100天时间。北极星就是一颗造父变星，光变周期接近4天。但北极星的亮度变化太小，肉眼很难察觉。莱维特的发现的意义在于她指出了较亮的造父变星比较暗的造父变星光变周期要长。更具体地说，造父变星的光变周期与其亮度之间存在确切的对应关系。例如，周期为5天的造父变星，其亮度是周期为11小时的造父变星的10倍。

莱维特是在研究小麦哲伦星云（Small Magellanic Cloud，SMC）中数百颗恒星时发现这一现象的。小麦哲伦星云是一个与银河系有关的恒星系统。她不知道小麦哲伦星云离我们有多远，但这并不重要，因为对我们来说，该星云中所有恒星与地球的距离基本上都是相同的。所以，无须担心距离不一致引起亮度上的差异，可以直接比较它们的亮度。1913年，丹恩·伊格纳·赫茨普龙（Dane Ejnar Hertzprung）采用几何法测量了离地球最近的13颗造父变星的距离，并利用这些观察资料，结合莱维特的数据，计算出了一个假想的、光变周期恰好为 1 天的标准造父变星的真实亮度。有了赫茨普龙的距离基准（周光关系），就有了测出其他任何造父变星的距离的可能：根据造父变星的光变周期与赫茨普龙的距离基准（周光关系），就可以计算出造父变星的真实亮度（绝对星等），然后[8]将结果与这颗造父变星在天空中的视亮度（视星等）进行比较，进而得出距离。造父变星的视亮度越暗，距离就越远。依据这个基准（周光关系），小麦哲伦星云至少位于 10 千秒差距之外。尽管后来根据更精确的观测数据和对恒星消光的理解，人们对赫茨普龙当初的基准进行了修正，但在 1913 年，相比于卡普廷的估算来说，这个值标志着整个银河系的大小（整个宇宙！）已经有了大幅增加，小麦哲伦星云离我们是如此遥远。

正是沙普利本人，在创建了自己的造父变星光变基准之后，利用造父变星技术描绘了银河系的大小和形状。这是他对这场大辩论的核心贡献。

沙普利成功测量银河系大小的关键在于，他能够使用变星来测量被称为**球状星团**（globular cluster）的恒星系统的距离。顾名思义，球状星团是球状的恒星系统。银河系平面的两侧都可以看到球状星团。球状星团可能包含成千上万颗独立的恒星，而在星团的中心有多达1000颗恒星聚集在1立方秒差距的空间中。这与银河系中我们所在的区域完全不同，太阳周围1秒差距的距离内没有任何恒星。通过测量到球状星团的距离，沙普利发现这些星团在太空中呈球形分布，球心在银河系光带中部朝向人马座方向的一个点上，距离我们有数千秒差距。这个点标记了银河系的中心，说明太阳系位于银河系非常边缘的地方（图1）。1920年，沙普利估算出了我们的星系大约有30万光年（近100千秒差距）宽，太阳距离银河系中心大约有6万光年[1]（近20千秒差距）。他在华盛顿会议上将之表述如下：

9

> 星团理论得出的一个结论是，太阳离银河系中心非常远。我们似乎靠近一个巨型星团或星云的中心，但这片星云离银河系中心至少有6万光年。

---

1 现在人们已测得，太阳距银河系中心有2.6万光年（8千秒差距）。

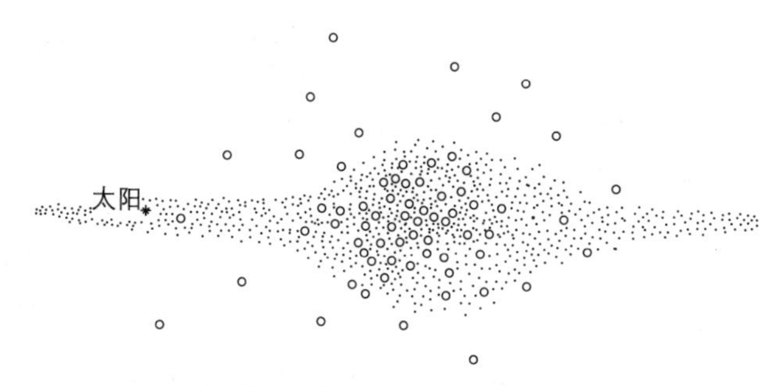

图 1　球状星团（圆形）只分布在天空的一侧，意味着太阳远离银河系的中心

从这张图上，沙普利及其他和他志同道合的天文学家看到，旋涡星云似乎不可能是像银河系那样的独立星系。他们的理由很简单：天空中物体的大小不仅取决于其真实大小，也取决于其与观察者的距离。实际生活中有很多这样的例子，比如一头站在田地另一边的牛，远看就像拿在手中的玩具牛一样大。如果旋涡星云的直径在 30 万光年左右，那它们在天空中呈现的视角如此微小，显然意味着它与们我们有数百万光年的距离，这也太大了。沙普利辩称，旋涡星云要么是银河系内的造星系统，要么至多是银河系的小卫星星云。与银河系大陆相比，它们是岛屿。“我宁愿相信，”他说，“那里面根本没有恒星，只是些云状物质罢了。”

10

他还备有另外一个撒手锏。荷兰天文学家阿德里安·范·马南（Adriaan van Maanen）恰巧是沙普利的好朋友。他声称自己通过对比相隔几年拍摄的照片，测量了几个旋涡星云的旋转，发现了非常小的旋转。他说，测量到 M101 星云转过了 0.02 弧秒，大约是从地球上观看月球视角大小的 0.001%。这样的旋转可以换算成星云上任意位置处与旋转中心的距离成正比的线速度。当然，这取决于旋转物体的实际大小。如果这些旋涡星云的大小与银河系相当，范・马南的测量结果就意味着某些位置的速度几乎与光速相同，甚至比光速更快。如果他的说法是对的，那么旋涡星云肯定很小，而且离我们较近。大多数天文学家很难相信范・马南能进行如此精确的测量。后来的研究表明，范・马南犯了错误，没有人确切知道是什么导致了这个错误，但在当时那场伟大的辩论中，是否相信他的数据是一个信任问题。沙普利信任他的朋友。在 1921 年发表的论文中，沙普利强调范・马南的研究结果对宇宙岛理论似乎是“致命的”，“那些明亮的旋涡星云，不可能像宇宙岛理论要求的那样遥远，那是不合理的”。

柯蒂斯不相信范・马南的测量结果，也不相信新鲜出炉的造父变星距离基准。在华盛顿的会议上，他概述了先前人们对银河系大小的各种估计，并故意将沙普利在 1915 年发表的银河系直径仅有 2 万光年的测算纳入其中。他得出结论："充分考虑传统观点，假定将银河系直径的上限定为3万光年，或许这的确太大了。"这一数字恰好是沙普利在1920年新估算值的1/10。柯蒂斯还表示，太阳虽然不是位于银河系的中心，但"相当接近"。对他来说，这些都是小事，只是在谈到他真正关心的话题之前简短地提了一下。柯蒂斯真正关心的是旋涡星云的本质以及它们到地球的距离。

柯蒂斯认为，就像银河系一样，旋涡星云是遥远的星系。柯蒂斯在论证过程中使用了两个关键事实。第一个事实是洛厄尔天文台维斯托·斯里弗（Vesto Slipher）的发现：到目前为止，大多数旋涡星云似乎正以极高的速度离我们而去。这是通过测量这些星云光谱中的谱线向红端移动的幅度，并与从附近恒星和地球上热物体发出的光的谱线对比而发现的。

来自任何热物体（包括太阳和恒星）的光，通过棱镜都可以形成彩虹图案，或者说光谱。每一种化学元素，比如氢、碳等，都会在光谱中产生一组特有的明亮谱线，就像超市商品上的条形码一样。当物体远离我们时，所有的谱线都向光谱的红端移动，移动的幅度取决于物体退行的速度，这就是著名的多普勒红移。同样，当物体向我们移动时，谱线条纹会向光谱的蓝端移动，这就是蓝移。在银河系中的恒星显示出既有红移，也有蓝移，对应速度从 0 到几十千米 / 秒不等。

在 20 世纪的第二个十年中，为测量旋涡星云微弱光谱中谱线的位置，摄影技术被推向了极致。直到 1912 年，斯里弗才获得了仙女座星系（也被称为 M31）的光谱图，现在我们知道它是离银河系最近的旋涡星系。他发现光谱向蓝端发生了位移，这表明该星云正以 300 千米 / 秒的速度冲向银河系。这是当时测量到的最高速度。到 1914 年，斯里弗已经获取了 15 个旋涡星云的类似光谱。其中只有 2 个星云（包括 M31）显示出蓝移，其他 13 个都显示出红移，并且 13 个中有 2 个显示出 1000 千米 / 秒以上的退行速度。到 1917 年，他获取的光谱中有 21 个是红移，但仍然只有 2 个是蓝移。即使到今天，仍然只有 2 个蓝移。无论旋涡星云的本质是什么，斯里弗测量到的速度表明，这些星云不可能是银河系的一部分，它们的运动速度太快，无法被引力束缚在银河系内。尽管 1920 年的时候，还没人清楚这些星云高速退行的原因，但柯蒂斯认为这就是证据，证明旋涡星云不仅与银河系毫不相关，而且旋涡星云就是实实在在的“宇宙岛”。

柯蒂斯使用的另外一个关键事实是观测到的恒星

突然耀发事件。这类恒星被称为新星（nova），源自拉丁文中的“新”一词，因为它们似乎是突然光芒四射地出现在以前没有恒星的地方。然而，现在一切都很清楚了，所有的新星都是恒星的爆发。爆发前这些恒星都非常普通，由于与背景相比太暗淡而不易被观测到。这是一种非常自然但相当罕见的恒星现象。

图 2　这是 NGC 4414，典型的圆盘星系，由哈勃太空望远镜上的第二代广域行星照相机（WFPC2[1]）拍摄

1　WFPC2，全名为 Wide Field and Planetary Camera 2，哈勃望远镜的第二代广域行星相机。

1920年，柯蒂斯指出："仅在过去的几年里，就在旋涡星云中发现了大约25颗新星，其中16颗位于仙女座星云中。与之相比，在银河系的全部历史记录中，只有30颗新星。"假设仙女座中的恒星不比银河系中的恒星更有可能成为新星，那么在仙女座星云中看到更多的新星说明其恒星数量比银河系多。大体上，从不同旋涡星云中看到的新星的视亮度（暗弱程度）应该与银河系中新星的视亮度是相同的。假如旋[13]涡星云的大小与柯蒂斯对银河系大小的估算相近的话，那么就意味着这些新星的距离遥不可及。

美中不足的是，1885年，也就是在仙女座星云被认定为旋涡星云的那十年里，一颗明亮的恒星在星云中突然耀发。这个新星的视亮度与银河系典型新星的视亮度大致相同。这意味着，要么仙女座星云其实是银河系的一部分，要么星云像柯蒂斯所说的那样遥远，而这颗恒星是某种超强新星，其亮度相当于10亿颗太阳的总和，比19世纪银河系观测到的任何新星都明亮得多。这对柯蒂斯来说是一个难题。[14]为回避问题，他提出可能有两种新星，一种比另一种亮得多。在当时的听众看来，这几乎是胡言乱语。现在，我们知道确实有如此壮观的恒星爆发，这就是超新星，短时间内可以像1000亿颗太阳一样明亮。事实上，超新星的亮度相当于星系中所有其他恒星的总和。

正如柯蒂斯在辩论中总结的那样：

> 似乎应当将旋涡星云视为星系，这样在其中观测到新星就显得很自然了。旋涡星云中的新星与银河系中的新星之间的相关性表明，旋涡星云的距离从50万光年（如仙女座星云）到1000万光年（更遥远的星云）不等……在这样的距离下，这些宇宙岛的大小与我们自己的恒星星系的大小等同。

在1921年发表的论文中，他进一步写道：

> 这些旋涡星系，作为河外星系，向我们展示了一个更大的宇宙，其间我们可以穿越1000万到1亿光年的距离。

1920年4月26日，在华盛顿，柯蒂斯和沙普利就宇宙的尺度问题展开了一场辩论，但没有人获胜。两名参与者都确信自己获胜，这本身就说明没有赢家。他们都有正确的一面，也都有错误的一面。最重要的是，沙普利对造父变星距离基准（周光关系）的信任是正确的，尽管那时造父变星距离基准还不完善；柯蒂斯对旋涡星云是独立星系的看法也是正确的；沙普利把太阳放在远离银河系中心的地方也是正确的。至于银河系的大小，目前最好的估计是直径约为 10 万光年，比柯蒂斯的估算值大 3 倍，是沙普利估算值的 1 / 3，所以他们的错误一样严重。而银河系的大小可以看作一个旋涡星系大小的平均值——我将在第 4 章讨论此问题。尽管这场伟大的辩论没有定论，但其提出的关键问题在 20 世纪 20 年代末之前得到了解决，这在很大程度上要归功于埃德温·哈勃（Edwin Hubble）所做出的贡献。

02

通往宇宙的“垫脚石”

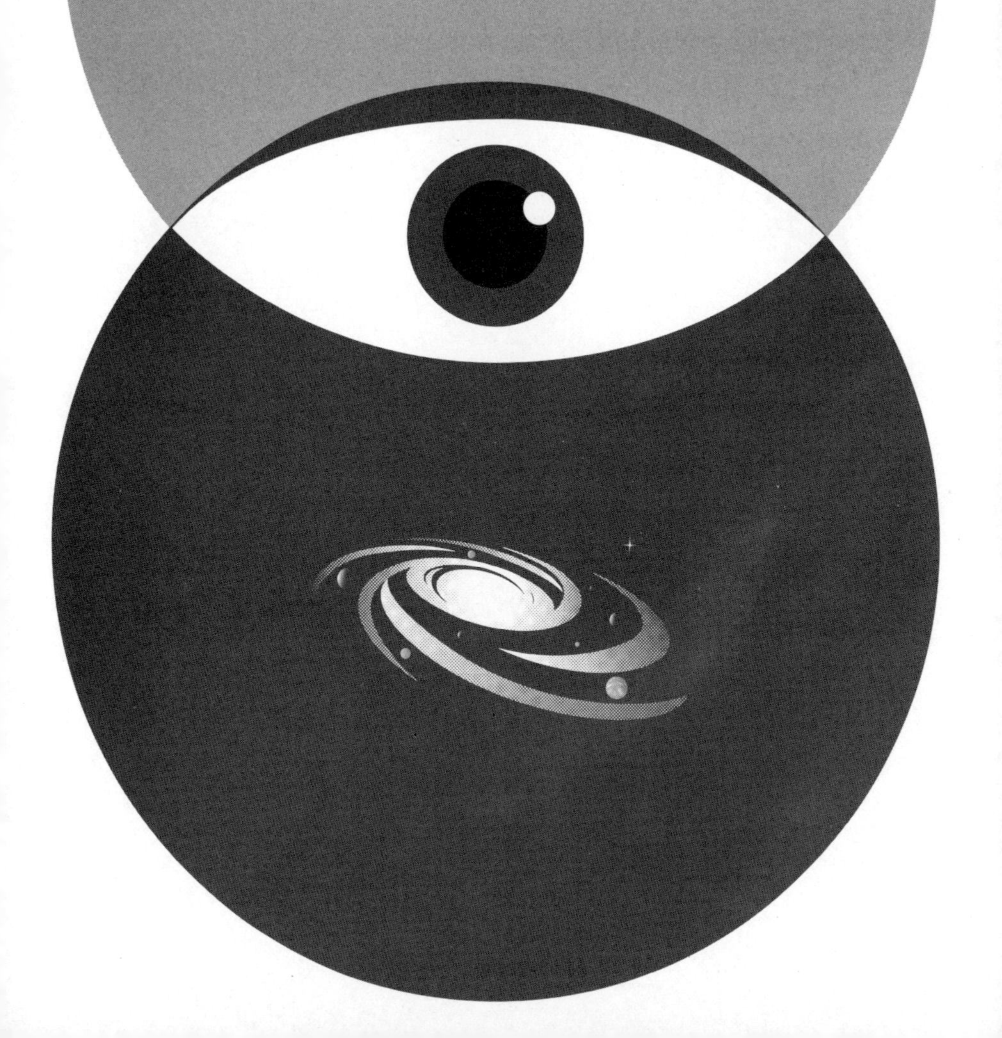

16

星系研究在 20 世纪 20 年代开始兴起，主要原因在于大型望远镜的发明和摄像技术的改进。有了这些先进设备的助力，我们就可以拍摄发光更微弱、距离更遥远的天体的清晰图像和光谱。分光光度法对发现旋涡星云光线的红移至关重要，而摄像技术则是发现造父变星周期与亮度间关系的关键技术。1918 年，在加利福尼亚州的威尔逊山上，一架口径 100 英寸（约 2.5 米）的望远镜[1]投入使用（图 3），它成为之后近 30 年里世界上最强大的望远镜。埃德温·哈勃正是用它和一系列精心设计的计算步骤，成功测量了宇宙中遥远星系与我们的距离。

1914—1917 年间，哈勃是芝加哥大学的一名博士研究生，而叶凯士天文台恰好隶属芝加哥大学。哈勃的天文学研究生涯正是从那一时期开始的。当时，他的研究项目是用一架 40 英寸（约 1 米）折射望远镜拍摄微弱星云的照片。当时，这是世界上最好的望远镜之一，也是有史以来人类建造的最大的折射望远镜。

1　下文中如无特殊说明，望远镜等设备的尺寸数据均指其口径。

图 3　威尔逊山上 100 英寸胡克望远镜的圆顶。哈勃就是用这架望远镜来测量其他星系的距离的

对于相同尺寸的望远镜，使用透镜的折射式望远镜一般比使用反光镜的反射式望远镜效果更好；但是反射式望远镜可以做得更大，原因是可以从背面支撑反光镜，同时又不会因此遮挡光线。按照观测计划，哈勃研究了星云的本质特征，并根据外观对星云进行了分类。到 1917 年时，这些观测已使他确信，星云一定远在银河系之外，特别是那些巨大的旋涡星云。

17

但在当时，这些想法并未得到更进一步发展。由于美国加入了第一次世界大战，1917 年 4 月，刚刚完成博士论文答辩的哈勃就自愿参军去了。哈勃曾在法国服役，并获得少校军衔，但从未参加过战斗。直到 1919 年 9 月，哈勃才返回威尔逊山天文台，成为该天文台正式的工作人员。在那里，他成了最早使用崭新的 100 英寸望远镜的人之一。抓住这次机会，哈勃将其博士论文发展成了一个完整的分类体系，并于 1923 年完成了此项工作。尽管哈勃一直用“星云”（nebulae）一词来指代那些虚幻缥缈的天体，但他确信这些星云位于银河系之外。这个想法很快就被证明是正确的。按照现在的说法，这些所谓的“星云”应该被称为“星系”（galaxy）。哈勃的早期工作揭示出的最重要的信息就是：太空中确实存在不同种类的星系，而巨大的旋涡星系只是这些天体中最醒目的一类而已。

哈勃指出，除了一小部分相对较小、形状不规则的星系，如小麦哲伦星云以及与其相似的大麦哲伦星云之外，所有星系都可以根据其形状来分类。椭圆星系是指那些从球形到拉长的透镜形等形状不一、没有明显内部结构的星系。旋涡星系是指那些内部有旋臂结构的星系，无论缠绕得松或紧，旋臂都是从星系中心向外展开的。也有的旋涡星系中央是一个棒状结构，旋臂从棒状结构的两端向外延展。哈勃认为，在演化的过程中，任何一种松散的旋涡星系，都将随着旋转，缠绕得越来越紧密，最终变成椭圆星系。在这一点上哈勃完全错了，但这并不影响他基于星系外

18 观的分类体系。目前人类已知的宇宙中，最大的星系正是椭圆星系。但是，也有比一些旋涡星系小的椭圆星系。另外，一些最初被认为是“旋涡”的星系，实际上是根本没有可识别旋臂结构的盘状恒星系统，因此，最好使用“**圆盘星系**”（disc galaxy）这个术语，它包含了那些有旋臂的星系。不过直到现在，许多天文学家在谈论基本上没有特征的圆盘星系时，也依然会使用“**旋涡星系**”（spiral galaxy）一词。

哈勃在威尔逊山天文台工作时，与沙普利有过短暂的交集。1921 年 3 月，沙普利离开威尔逊山天文台前往哈佛大学任职。当哈勃开始使用 100 英寸望远镜试图证实星云是独立星系时，反对他的更资深的天文学家已经离开了。不管怎么说，随着观测资料的不断丰富，在 20 世纪 20 年代初，宇宙岛理论开始得到部分天文学家的支持。那时，丹麦天文学家克努特·伦德马克（Knut Lundmark）访问了利克天文台和威尔逊山天文台，看到了名为 M33 的星云（星系）照片。这些照片足以说服他（虽然并未说服沙普利），图像上的那些颗粒能够表明星云是由恒星构成的。1923 年，有人在星云 NGC 6822 中发现了几颗变星，但它们过了一年才被确定为造父变星，那时，哈勃已经突破性地在仙女座星系（M31）中发现了造父变星。

实际上，哈勃并不是在寻找造父变星。1923 年秋，哈勃完成了星云分类体系后，又接手了柯蒂斯的主要工作之一，用 100 英寸望远镜开始了一系列拍摄观测，目的是在 M31 的一条旋臂中寻找新星。很快，也就是在那一年 10 月的第一周，哈勃发现了 3 个明亮的光点，在照片里看起来就像新星。由于这架 100 英寸望远镜已经运行了好几年，所以积累了一批拍摄档案，其中就包括对 M31 相同区域的观测照片。这些照片是由包括沙普利和弥尔顿·哈马逊（Milton Humason）在内的几个观测者拍摄的，哈马逊后来成为哈勃最亲密的合作伙伴。这些照片显示，哈勃暂定为新星的 3 个光点之一，实际上是造父变星，其光变周期稍长于 31 天。出乎预料的是，使用沙普利校准的造父变星距离标度进行计算，立即得出它距离地球有近 100 万光年（300 千秒差距）的惊人结果，这甚至比沙普利对银河系大小的估计的 3 倍还要大。出于更全面的考虑，他后来对整个距离标度进行了修正，其中部分原因就是考虑了恒星消光的影响。我们现在知道 M31 实际上位于 785 千秒差距远的地方，大约相当于银河系直径的 20 倍之外。1923 年，几乎就是从 M31 中发现这颗造父变星的同时，哈勃立刻意识到，这家伙确实是个星系！而且是一个位于银河系之外，多少有点像我们银河系的星系！

在接下来的几个月里，哈勃在M31中又发现了1颗造父变星和9颗新星，而且依据它们分别测出的距离大致相同。在其他星云中哈勃也发现了造父变星和新星。1925年1月1日，在华盛顿特区举行了美国天文学会和美国科学促进会的联合会议。哈勃把所有观测结果都写进了一篇论文中，并提交给了大会。哈勃并没有出席会议，而是由亨利·诺里斯·罗素（Henry Norris Russell）代他宣读了这篇论文。哈勃完全没有必要再进行其他说明，论文阐释了一切。与会者完全赞同他的观点，星云的本质特征最终得到了确定，宇宙比以前的估算大了很多，而银河系只是其中的一个“岛屿”——哈勃取得了完胜。其实，早在那次会议之前，哈勃就写信给沙普利，告诉了他这些发现。沙普利读这封信时，一个从1923年起就跟随沙普利攻读博士学位的研究生——后来的天文学家塞西莉亚·佩恩－加波施金（Cecilia Payne-Gaposchkin）碰巧就在沙普利的办公室里。“喏，”沙普利把信递给她时说，“这封信摧毁了我的宇宙。”虽然大辩论就此结束了，但哈勃使用了造父变星技术，这给沙普利的银河系模型及其学说——特别是否定太阳在我们的星系中位于中心地位这一点增加了分量。对沙普利来

20

说，这或许多少是一种安慰。

如果说沙普利的宇宙已经被摧毁了，那么新宇宙——哈勃的宇宙会是什么样子的呢？宇宙如此之大，即使是使用 100 英寸望远镜的哈勃，也只能拍摄那些距离非常近的星系中的造父变星的照片，更别说那些使用较小望远镜的观测者了。哈勃对测量宇宙尺度的想法非常着迷，甚至达到了痴迷的程度。这促使他积极寻找其他方法，以便测量造父变星技术无法测量的距离。到 20 世纪 20 年代中期，他开始了这项工作。

为此，哈勃收集了一系列“垫脚石”，借助这些“垫脚石”，观测者可以眺望越来越远的宇宙。首先是造父变星，造父变星的亮度只能用来测量离得最近的几个星系的距离，而且在哈勃太空望远镜问世之前只有几十个被发现。接下来是新星，新星比造父变星稍亮一些，可观测的距离更远。在根据造父变星确定了 M31 的距离之后，哈勃就用这个距离校准了 M31 中新星的亮度。这样一来，新星就成了距离测量的新标尺。然后，哈勃先假设所有的新星都具有相同的固有亮度，接下来，利用新星测量更遥远星系的距离。100 英寸望远镜及后来的望远镜拥有强大的分辨力，这对哈勃来说，意味着其他测距技术也变得可行。比

21

如，星系中最亮的恒星比造父变星亮得多，也可以充当示距天体。这里他假设所有星系中最亮恒星的亮度都一样，原因是恒星的亮度肯定有上限。哈勃还利用遥远星系中的球状星团，猜测每个星系中最亮的球状星团一定具有大致相同的固有亮度。超新星的秘密一经破解，也以同样的方式被加入破解宇宙秘密的“垫脚石”行列。

另外，哈勃还基于整个星系的亮度，以及星系在天空中的表观（角度）大小，进行更粗略更迅速的距离估计。如果每个旋涡星系都和 M31 一样明亮，都和 M31 一样大，那么只需将这些星系与 M31 进行比较，就很容易测量它们的距离。不幸的是，事情远不是如此简单，哈勃自己也心知肚明。但由于没有更好的方法，所以他将一些看起来非常相似的星系进行比较，为的是至少能得到一点有关于距离的指引。

这些技术都称不上十全十美，但只要有可能，哈勃还是尽可能多地将这些技术应用到每个星系上，希望能借此消除一些误差和不确定性。然而，这一切都需要时间。1926 年，哈勃着手绘制银河系周围的星系分布图，沿着斯里弗等人得到的红移数据中的线索，借助充足的数据，再加上周密的思考，哈勃的研究终于取得了巨大的飞跃。

1925 年，哈勃在对星系进行光谱分析后发现，其中发生红移的有 39 个，发生蓝移的只有 2 个。事实上，斯里弗才是第一个测量这些星系红移的人。哈勃测量的星系中只有 4 个是斯里弗没测过的。但斯里弗受望远镜性能的限制，没能继续测量。斯里弗使用的是洛厄尔天文台的一架 24 英寸（约 60 厘米）的折射

23

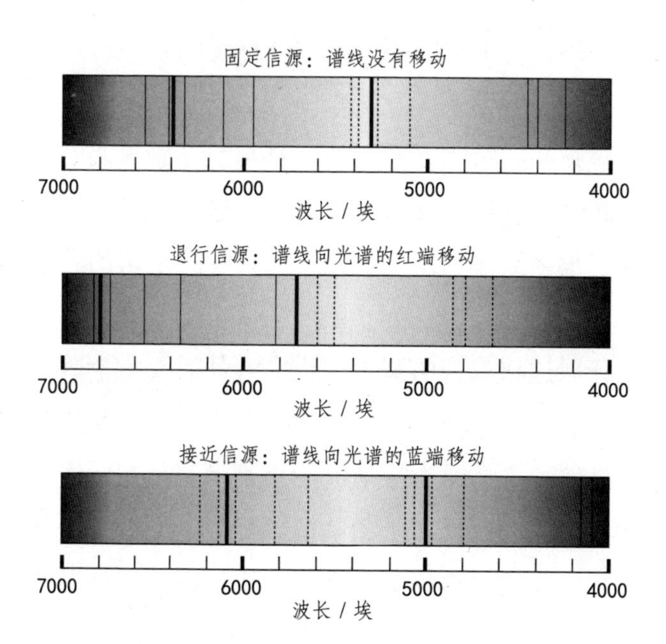

图 4　恒星相对于观测者的运动速度和方向决定了谱线移动的幅度。当辐射天体远离观察者时，发出的波被“拉伸”，波长变长，谱线向光谱的红端移动。如果辐射天体靠近，波长就会被压缩，谱线就会向光谱的蓝端移动。红移可以用来计算物体的退行速度

望远镜，总计测出了 43 个星系的红移。从这些数据可以隐约看出，红移越大，星系的距离就越遥远。其实，这一点早就有人注意到了，只不过是哈勃拥有得天独厚的天文学家身份，坐拥当时世界上最强大的望远镜，因而最有机会证实这一点罢了。他就是那个出现在正确的时间、正确的地点的人。哈勃想要找出星系距离和红移之间的精确关系，这是他逻辑链条上的最后一环。有了这一环，通过红移量测量宇宙大尺度的距离就成为可能。

1926 年，哈勃开始寻找红移与星系距离之间的联系，此时，他将所有精力都集中于这一工作。他已经获取了很多天体的距离数据，并将在接下来的几年里获取更多。但是 100 英寸望远镜从未用于红移测量，为了开展这项艰苦的测量工作，哈勃需要一个既有能力又愿意付出的同事。哈勃选择了出色的观测员哈马逊。显然，这是因为哈马逊的级别没有他高，因此外界能清楚地看到谁才是团队的领导者。经过一番艰苦努力后，他们改造好了 100 英寸望远镜，哈马逊特意选择了一个特别暗淡的星系作为第一个红移测量对象。因为这个星系太暗淡了，所以斯里弗从未用这种方法研究过它。哈马逊得到的红移速度相当于 3000 千米 / 秒，是斯里弗得到的最大测量值的两倍多。哈勃和哈马逊的合作伙伴关系正式开始了。

到 1929 年，哈勃确信已经发现了红移和星系距离之间的关系。而且这正是他所希望的最简单的关系——红移与星系距离成比例。或者，换句话说，星系距离与红移成正比，对哈勃而言这才是最重要的。如果一个星系的红移是另一个星系的两倍，那么其与我们的距离也是另外那个星系的两倍。二人合作取得的第一批成果发表于 1929 年。其中，哈勃提供了 24 个已知红移和已知距离的星系的数据，并根据这些数据计算出“红移 – 距离关系”的比例常数为 525 千米 /（秒 · 百万秒差距），写作 km /（s · Mpc）。也就是说，一个具有相当于 525 千米 / 秒红移速度的星系，距离地球有 100 万秒差距（325 万光年），其他星系依此类推。哈勃选择这个特定的数字就像其他任何事情一样，看起来都是一厢情愿的，因为有限的数据量并不足以证明所用数字的准确性。但在 1931 年，哈勃和哈马逊共同发表了一篇论文，又加进 50 个红移星系数据，强化了这个结果（图 5）。其中，最大的红移量相当于 2 万千米 / 秒，与哈勃 3 年前得到的数字更为接近。很明显，早在 1929 年哈勃就获得了其中一些数据，但出于某种原因，他选择将这些数据秘而不宣。

24

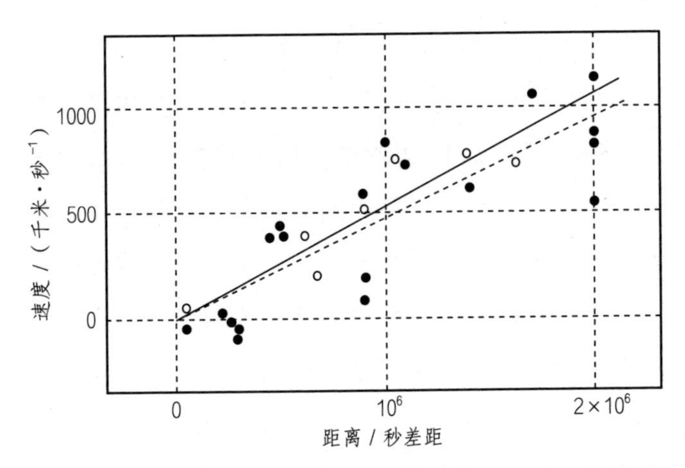

图 5　基于 1929 年发表的数据，哈勃最初对“红移 – 距离关系”的解释还说得过去；到 1931 年，他与哈马逊合作，给出了一个更有说服力的解释

哈勃既不知道也不关心为什么红移 – 距离关系会存在，他甚至没有提到这种关系意味着红移的星系正在退行。虽然按照惯例，红移通常以千米 / 秒为单位，但除了在太空中运动产生红移之外，还有其他原因可以产生红移（例如强引力场）。20 世纪 30 年代，哈勃谨慎地猜想，可能存在未知的因素在起作用。他在《星云王国》（*The Realm of the Nebulae*）一书中写道：

为了方便起见，红移可以用速度单位来表示。红移的变化与速度变化一样，并且可以非常简洁地用熟悉的尺度表示，不管最终的解释是什么。“表观速度”（apparent velocity）一词的使用要谨慎，前面的形容词有重要的含义，并不像在一般用法中那样可以省略。

不管红移－距离关系的起源是什么，哈勃确实证明了它就是测量宇宙尺度的终极工具，其比例常数被称为哈勃常数，用 H 表示。自 1931 年以来，所有对银河系以外天体距离测量的目的只有一个，那就是校准哈勃常数。所有这些研究，对于理解星系及其在宇宙中的位置都具有重大意义。但是，在此之前，似乎应该首先梳理一下对银河系的理解。银河系——这个普普通通的圆盘星系，是人类孤悬在太空中的唯一家园。

26

03

银河系

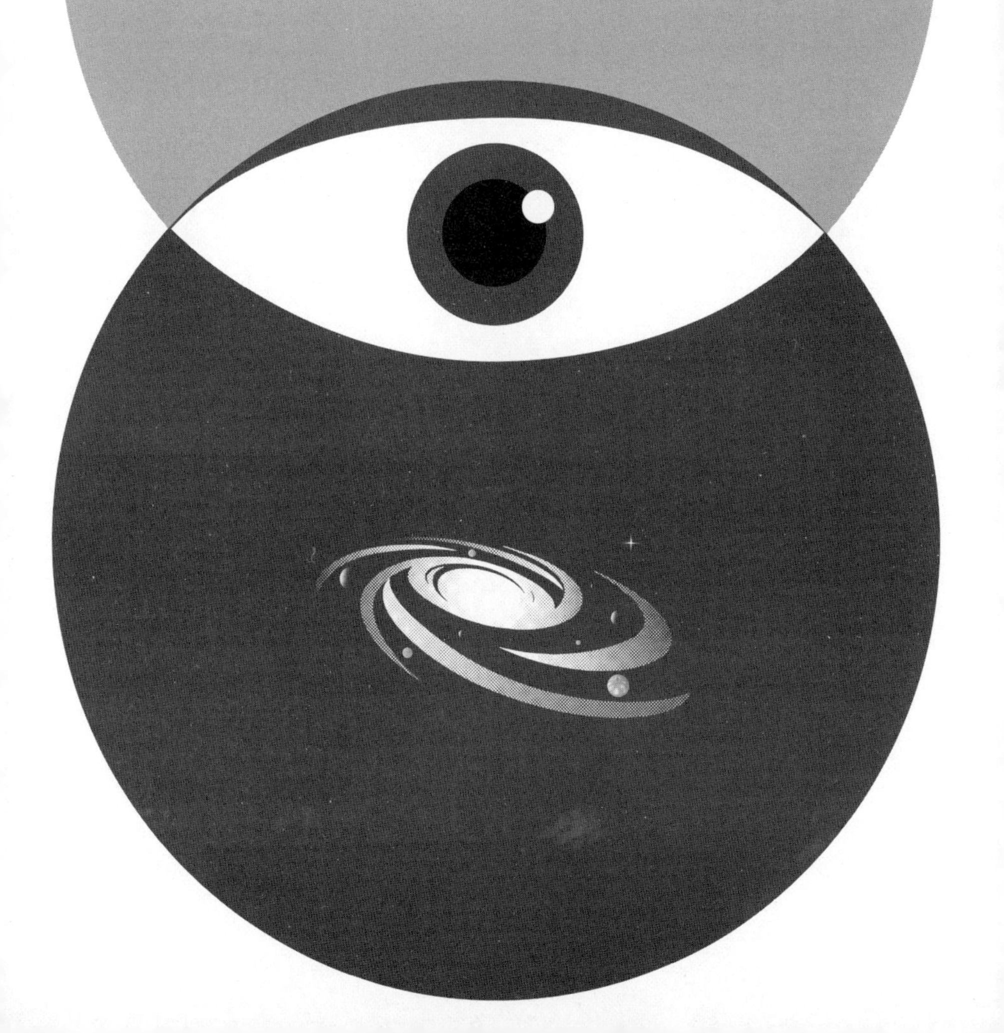

自20世纪20年代以来，人类对银河系的了解迅速增加，这主要得益于科学技术的不断进步和观测水平的持续提高。在观测手段方面，除了有包括哈勃太空望远镜在内的更强大的光学望远镜之外，还有射电望远镜、红外望远镜、X射线探测器，以及卫星上携带的各种其他设备。相比于照片和哈勃那代人使用的光谱仪器，现在使用的灵敏的电子探测器能够获得更多较暗天体的信息。同时，功能强大的现代计算机也使理论预测与观测结果的比较比哈勃时代容易得多。 [27]

从20世纪20年代至今，与银河系相关的最重大的发现莫过于神秘的暗物质。很明显，从整个星系的旋转方式来看，这个明亮的星系盘受控于一个大致呈球形的暗物质晕的引力。它涉及的暗物质总量，大约是哈勃所能想象的星系质量总和的7倍。而其他所有发光的恒星只占银河系总质量的一小部分。暗物质的发现，深刻影响着人类对整个宇宙的认识，因为普通物质和暗物质的比例似乎同样适用于整个宇宙。彼 [28] 得·科尔斯（Peter Coles）在《宇宙学》（*Cosmology: A Very Short Introduction*）中阐述了宇宙学的含义：除了暗物质的存在问题之外，最重要的一点是暗物质不是简单的寒冷气体或尘埃，也不像太阳、恒星和我们那样，由原子构成，而是完全由别的东西构成的。由于没有人知道它到底是什么，因此它被简单地称为**冷暗物质**（Cold Dark Matter，CDM）。

太阳是一颗典型的恒星。尽管恒星的质量有大有小，但都以同样的方式运作——通过核聚变将内部的轻元素（尤其是氢）转化为更重的元素（尤其是氦），释放出维持恒星发光的能量。总的来说，据估计，银河系中有几千亿颗（至少3000亿颗）恒星，散布在一个直径约为28千秒差距，也就是9万多光年的星系盘上。这个星系盘的大小很难确定，就如同一个人置身于森林之中，就很难对森林的大小做出判断一样。所以，这些数字通常四舍五入到30千秒差距或10万光年。在银河系星系盘的中央，大量恒星聚集呈球形[1]，如果从侧面看，就像两个煎蛋背对背地粘在一起。整个星系盘被古老恒星和球状星团构成的球状光晕所环绕，光晕中包含了银河系中最古老的恒星。我们已知的球状星团有近150个，至少还有50个是看不见的，因为银河系明亮的光带介于我们和那些

1 星系盘隆起的球状部分叫“核球”，四周叫“银盘”。

星团之间，遮挡了我们的视线。

天文学家可以利用多普勒效应研究恒星在太空中的运动方式。总的来说，如果一颗恒星在退行，也就是远离我们而去，其光谱线就会向光谱的红端移动；如果恒星在向我们奔来，光谱线就会向蓝端移动。这种效应的大小揭示了恒星的径向速度。这与运动物体发出的声音十分类似，比如救护车上的汽笛声——当救护车远离你时音调变低，而接近你时音调会变高。早在 1842 年克里斯蒂安·多普勒（Christian Doppler）就预测到了这种效应。他让号手在火车上吹奏一个固定音符，当火车经过时，对其进行了测量。从表面上看，这种效应类似于星系的红移，但宇宙学上的红
29
移[1]并不是天体在空间中运动所引起的，不属于多普勒效应。

太阳的位置，大约是从银河系中心算起，到星系盘边缘的 2 / 3 处（略小于 10 千秒差距）。像星系盘中的其他恒星一样，太阳绕着星系中心，以大约 250 千米 / 秒的速度在一个大致呈圆形的轨道上运行，绕银河系转一圈需要近 2.5 亿年。恒星的年龄，可以通
30
过将它们的整体外观（尤其是它们的颜色和亮度）与

1 宇宙学红移是指当光穿过一段空间时，空间自身尺度的变化导致的光线波长变化。

理论模型进行比较来确定。这些理论模型描述了恒星随核燃料的消耗会发生什么变化。例如，可以通过测量岩石和陨石的放射性，推断出太阳系的年龄来证实这种理论。太阳和太阳系已经存在了大约45亿年，围绕银河系中心转了近20圈。自从第一个人类——也就是现代智人（Homo sapiens）出现在地球上以来，太阳系只围绕银河系中心转过了不到千分之一圈。银河系中最古老的恒星有130亿年多一点儿的历史，这约为太阳年龄的3倍。

除去中央核球部分，银盘的厚度只有约300秒差距[1]，也就是大约1000光年厚。太阳系只比星系盘的平面中心高6～7秒差距。俯视整个银河系，星系盘看上去并不像煎蛋，而是像中心被一条8～9千秒差距长的棒形结构横穿的中央核球。有4个缠绕得相当紧密的可见旋臂从星系的中部盘旋而出。就像其他圆盘星系那样，旋臂之所以明亮，是因为其中包含了许多处于初期的炽热的年轻恒星，这些恒星又大又亮。恒星的质量越大，就越需要猛烈地燃烧核燃料以抵消引力带来的压缩效应，消耗燃料的速度也就越快。旋臂是恒星形成的场所。较小的、寿命较

1　原文为300千秒差距，明显有误。一般统计数据为600秒差距或2000光年。

长的恒星，比如太阳，就形成于旋臂中，但不会那么明亮。太阳系目前位于一个被称为猎户座旋臂的分支旋臂中，这个臂也称为“近域旋臂”。这个分支在两个主旋臂之间形成桥梁。沙普利认为太阳系处于一个局部的恒星集中区的想法是正确的。

主要存在于旋臂和银河系平面（以及其他星系盘）中的年轻恒星被称为星族Ⅰ星[1]，这些恒星内部包含了从前几代恒星中散布出来的物质，其中的重元素就是构成地球等行星的主要成分。太阳就是一颗星族Ⅰ星。在星系光晕、球状星团和核球处发现的较老恒星被称为星族Ⅱ星，它们比星族Ⅰ星更红。星族Ⅱ星都比较古老，形成时银河系还很年轻。这些恒星主要由宇宙诞生时大爆炸产生的原始氢和氦构成。在星族Ⅰ星和地球中存在的较重元素，是在前几代恒星中形成的。椭圆星系主要由星族Ⅱ星构成。

32

如果不考虑其他因素，随着恒星围绕星系的运动，像银河系这样的星系中所看到的旋涡状图案，大约在10亿年之内就会被抹去。旋涡图案之所以能够持续存在，是因为有大量气体和尘埃云团围绕着银河系做轨道运动。当这些云团穿过旋臂时受到挤压，产

1 也称第一星族。

生类似于音爆那样的冲击波，就会导致新的恒星接连不断地形成。这些年轻的恒星，就是环绕星系运行的冲击波最明显可见的特征。

这种现象可以拿繁忙的高速公路上发生的交通拥堵做类比。当一辆缓慢行驶的重型车占据了内侧车道，后面速度快的车辆想超车就要转入外侧车道，最后造成重型车一侧车道的交通拥堵。尽管车辆不断从拥堵的前端驶出，但又会有新的车辆从后面补充进来，整个拥堵车队就会缓慢且稳定地向前移动。同样，旋臂以恒定的速度绕着银河系运动，但是新的气体和尘埃云团会不断加入其中，被压缩，然后继续前行。受挤压影响，其中一些云团自发地形成了恒星。

尽管是自发形成，但显然，这一过程并不是非常高效。因为若是真的高效，时至今日，银河系所包含的所有气体和尘埃，早就都形成恒星了。事实上，在银河系中，每年只有几个太阳质量的物质转化成新恒星。这大致与老恒星死亡时抛回太空的物质的质量相等。因此在圆盘星系中，恒星的诞生、演化和消亡的过程能够持续数十亿年。这也意味着，在银河系形成并稳定下来之前的短暂时间内，一定有很多恒星集中诞生。这叫恒星爆发（图 6），它非常壮观，在其他星系中也能看到。

图 6　恒星爆发的星系 M82。这是一张综合了来自 WFPC2 和美国基特峰天文台 3.5 米望远镜所得数据的合成图像

由气体和尘埃组成的云团很难坍缩成一颗恒星（或几颗恒星），原因有二。首先，所有的云团都在旋转。即使只是轻微旋转，当坍缩发生时，因角动量守恒旋转也会迅速加快，从而抵消了引力。因此，云团必以某种方式保持分散状态。其次，在云团坍

缩时，释放出来的引力势能使云团变热，除非这些热量能被辐射出去，否则不可能进一步收缩。在恒星形成的过程中，角动量的问题是由云团分解成几颗恒星来解决的。这样，云团的角动量就转化为绕彼此旋转的恒星的角动量。平均而言，每100个新诞生的恒星系统中，大约有60个是双星系统，40个是三星系统。像太阳这样的孤立恒星，是后来才从三星系统中甩出来的。因为云团中含有一氧化碳等分子，热量问题也可以解决。这些分子能够在光谱中的红外部分吸收或辐射热量。尽管如此，恒星的形成仍然是艰难的过程，这很奇怪，因为恒星竟然真通过这样的过程形成了。

恒星的形成始于巨大的气体复合体，这些气体可能是横跨1千秒差距，包含1000万个太阳质量的物质。其中独立的云团可能只有几十秒差距宽，包含几十万个太阳质量的物质。最初导致云团坍缩的压力很可能由一颗大质量恒星的爆发而产生，即超新星。坍缩云团内的湍流，引导了直径约为0.2光年的核心形成，其质量约为太阳的70%。但是整个云团的质量中以这种方式转换成核心的只有百分之几。当一颗恒星开始形成时，首先达到形成恒星所必需的密度的内核更小，质量大约只有太阳质量的

34

千分之一。恒星的其他质量是依靠吸积周围的尘埃云团物质积累起来的。这些物质离恒星很近，足以被引力吸进去并落在核心上，所以恒星的最终质量取决于附近有多少物质。一旦恒星开始发光，其辐射就会把周围剩余的物质吹散。

整个过程很快就结束了。一个云团坍缩形成一簇恒星，炽热的年轻恒星将剩余物质吹走，这一切都发生在大约 1000 万年之内。这个过程的晚期情形可以在附近的猎户座星云中看到（图 7）。但是一些新形成的年轻恒星质量比太阳大得多，并且会很快耗尽核燃料。这些恒星将以超新星爆发的形式结束生命，它们产生的强大冲击波横扫周围的星际物质，并触发气体和尘埃云团的坍缩。这似乎是一个自发的过程，在负反馈的帮助下，像银河系这样的星系可以保持稳定状态。在某一代恒星中或某一位置上，如果形成的恒星数量超过平均水平，那么来自这些恒星的能量会将气体和尘埃推散到更大的区域，从而减少下一代恒星的数量；但是，如果形成的恒星数量较少，就会有较多的气体和尘埃留下来，在接下来的循环中，形成更多的恒星。回归平均水平是自然趋势。能形成超新星的恒星，在几百万年内就会燃烧殆尽，这与太阳已有的 45 亿年的寿命相比，显得非常短暂。所有这些活动都发生在旋臂之内，有助于星系保持旋涡状。

图 7　猎户座的恒星形成区。由斯皮策太空望远镜（Spitzer Space Telescope）拍摄的红外图像

在我们星系的中心区域，即整个旋涡形状的中心，而不仅仅是星系盘的中心，存在着一个大质量的黑洞，其质量大约是太阳的460万倍。我们将在第7章看到，这样的黑洞是星系存在的关键。

大多数的黑洞要比上述黑洞小得多，一般质量只有太阳的几倍。根据广义相对论的规定，如果一颗恒星在其生命末期的质量超过今天太阳质量的3倍，就会形成这样的黑洞。这样的恒星，由于燃料耗尽，内部不再产生热量，其残骸会因无法承受自身的重力而坍缩，最终收缩成一个体积为零的奇点。原子及其组成粒子——质子、中子和电子，都在这个过程中被粉碎，不复存在。几乎可以肯定的是，在到达奇点之前，广义相对论就已经失效了。但在远未到达奇点之前，坍缩体的引力变得极其强大，以至于任何东西都无法逃脱，光也不例外。这就是黑洞这一名称的由来。换个说法便是：黑洞的逃逸速度超过了光速。既然没有东西能比光速更快，那么也就没有东西能从黑洞中逃脱。

任何物体只要压缩得足够小，都会变成黑洞。事实上，任何质量的物体，都有一个临界半径，叫作史瓦西半径。对太阳来说，史瓦西半径略小于3千米；对地球来说，史瓦西半径只有不到1厘米。总而言之，不管什么样的物体，只要全部质量被压缩到它的史瓦西半径内，就会变成黑洞。

尽管黑洞本身是不可见的，但黑洞的引力会对周边环境产生非常强烈的影响，这种现象不难被探测到。有些恒星级黑洞与普通恒星邻近，两者携手做轨道运行，形成双星系统。在这种情况下，通过研究黑洞引力对伴星轨道的直接影响，可以计算出黑洞的质量。另外，从伴星拉出的物质呈漏斗状流入黑洞的“喉咙”。在下落过程中，物质流中的粒子不断加速并相互碰撞，温度迅速升高，甚至能发射出 X 射线。

所有这些黑洞都与物质被压缩到极高的密度有关，但位于银河系中心的黑洞是另一种怪兽。奇怪之处在于，早在阿尔伯特·爱因斯坦（Albert Einstein）提出广义相对论之前，这样的超大质量黑洞就引起了理论物理学家的好奇。1783 年，英国皇家学会的研究员约翰·米歇尔（John Michell）指出，根据牛顿的引力理论，直径是太阳直径 500 倍的天体（大约与太阳系直径一样大），如果其密度与太阳相同，其逃逸速度会比光速还快。当时，米歇尔并没有使用“逃逸速度”（escape velocity）这个词。但在现代语言中，“逃逸速度”就是他所要表达的意思。当然，爱因斯坦的理论也做出了相同预测。不过，米歇尔的黑洞根本谈不上超密度，因为太阳的总体密度只有水的 1.5 倍左右。1796 年，法国人皮埃尔·拉普拉斯（Pierre Laplace）也得出了同样的结论。他说，虽然这些暗物体无法直接被看到，但“如果有任何其他发光物体碰巧围绕它旋转，也许我们就可以根据这些旋转体的运动规律，推断出中心物体的存在”。两个世纪后，位于银河系中心的黑洞正是通过这种方式被发现的。

银河系的中心位置与人马座在天空中的方向相同，但要远得多。“人马座”这个名字，是古人根据恒星组合成的图案命名的。这些恒星看起来很明亮，仅仅是因为它们离我们很近。如今，天文学家们仍在使用这些名字来表示星体位于天空的哪一部分或哪个方向，这就是为什么 M31 也被称为仙女座星云（或仙女座星系）的原因。要知道，M31 比仙女座的恒星要远几百万光年，而且与仙女座毫无瓜葛。同样，银河系中心一个强大的射电波源被称为**人马座 A**（Sagittarius A，缩写为 Sgr A），尽管它与人马座的恒星没有任何关系。

37

只有使用射电望远镜那样不依赖可见光的设备，才有可能观察到银河系的中心，因为在银河系的平面上有大量的尘埃挡住了可见光。这些尘埃导致了星际消光，也严重影响了早期人们对距离标度的测定。不过，这些尘埃也是形成新一代恒星的原材料。较长的波长更容易穿透尘埃，这就是为什么日落是红色的。短波长（蓝色）的光被大气中的尘埃散射到视线之外，较长波长的红光才能进入人们的眼帘。因此人类对银河系中心的了解，也就主要是基于红外线和射电观测。

更详细的研究表明，人马座 A 实际上是由 3 个相互靠近的天体组成的。一个是与超新星残骸相关的膨胀气泡；一个是热的电离氢气区域；第三个被称为人马座 A*（简称 Sgr A*），位于银河系的正中心。

在人马座 A* 周围，天体运动非常活跃。红外观测显示，这个区域中的恒星密度非常高，在 1 立方秒差距的空间内，聚集了 2000 万颗像太阳那样的恒星。这些恒星之间的平均距离，仅是地球到太阳距离的 1000 倍。由于距离太近，每隔 100 万年左右就会发生碰撞。这个核球周围环绕着一个巨大的气体尘埃环，向外延伸 1.5～8 秒差距（4.5～25 光年），带有近期爆炸产生的冲击波痕迹，X 射线和能量更高的 γ 射线从中心区域射出。

38

对所有高科技观测设备而言，观测到拉普拉斯设想的那种场景，才是黑洞存在的最佳证据。夏威夷莫纳克亚天文台（Mauna Kea Observatory）使用 10 米反射式望远镜在红外线波段进行观测，获得了靠近银河系中心的 20 颗恒星移动速度的观测数据。数据显示，这些恒星正以高达 9000 千米 / 秒的速度绕着银河系中心旋转。这些恒星运动得实在太快了，在连续几年拍摄的一组照片中，只需间隔几个月，这些恒星的位置就会发生变化。要知道，这些恒星远在 10 千秒差距之外。把这些照片拼接在一起，就能像电影那样展示出这些恒星靠近星系中心的轨道运动。它们的轨道运动说明，这些恒星被一个质量是太阳质量的 460 万倍的物体所控制。由于这些恒星轨道所界定的区域并不比地球绕太阳公转轨道占据的区域大，所以该物体绝对是一个超大质量黑洞。

相较而言，这个黑洞现在还比较安静，因为它已经把附近的所有物质都吞没了。现在可供探测的黑洞活动，仅是从环状吸积盘上落入黑洞的少量物质流造成的。物质落入黑洞时会释放出引力势能，这让辐射可被观测。按照目前的活动水平，这个黑洞每年要“吞噬”的物质总质量仅相当于太阳质量的 1% 左右。在银河系形成的早期，黑洞周围区域的气体和尘埃还没被清除时，情况肯定有所不同。稍后会讨论这个问题，但很明显，超大质量黑洞是星系生长的种子。

39 可以肯定，到目前为止所描述的银河系的有序结构，包括核球、银盘和光晕等，并不是星系的全部。恒星从星系中心向外移动的方式也包含着星系演化的丰富信息。在观察单个恒星的构成及运动方式时，天文学家发现，以银河系里的众多恒星为背景，可以分辨出多个形状细长的星流。每个星流里构成恒星的物质相似、运动方向相同，但它们与银河系里大部分恒星的构成成分不同，运动方向也与背景恒星呈一定的角度。

现在已经确定了 70 个这样的星流（采用哪个数字取决于你相信哪个证据），一定还有更多星流尚待发现。星流的质量范围从几千到 1 亿个太阳质量不等，长度从 2 万～100 万光年不等。通常，逆着星流运动方向观察会发现，这些恒星系统隐约与球状星团或围绕银河系运行的 20 个左右的小星系有联系。这些小星系绕着银河系运行，就像卫星绕着行星运行一样。从地球的角度来看，这些恒星径迹中最壮观的属于**人马座星流**（Sagittarius star stream），其弯曲的径迹跨度超过 100 万光年，两端连接着银河系与所谓的**人马座矮椭圆星系**（Sagittarius dwarf elliptical galaxy）。另一个星流是在室女座（Virgo）方向上发现的，因此被称为室女座星流，其运动方向几乎与银河系平面垂直，与另一个矮星系相通。

此迹象解释了星流的起源。那些离我们星系太近的小星系，会在银河系强大的潮汐力的作用下解体，形成像尾巴一样长长的星流。小星系的残余部分连接在星流的末端，继续沿环绕银河系的轨道运行。人马座矮星系处于这一过程的最后阶段，时至今日，已经几乎看不出它曾是一个结构紧密的恒星群。最终，它除了星流什么也不会留下，而星流最终也将融入银河系。

这清晰地表明，银河系是通过星系间的同类相食，吞噬较小的邻居，才成长到了现在的规模。利用强大的统计学技术，天文学家甚至能够从现在对星流运动的观测中，反向重建前卫星星系的结构，就像古生物学家从一些化石残骸中重建恐龙的外观那样。这些星流的轨道形状就像蛋糕上的糖霜，这告诉我们，暗物质从银河系中延伸出去的晕是球状的，而不是椭球形的。

40

然而，这些星系间的相互作用并不局限于大星系吞并小星系这一种情况。斯里弗发现，仙女座星系发出的光出现了蓝移，折算成的速度接近 100 千米 / 秒（近 36 万千米 / 小时）。正如哈勃所意识到的，它之所以没有出现红移，是因为宇宙学的红移并不是天体在空间中运动引起的。在仙女座星系那么近的地方，宇宙红移量很小，等效速度项的大小不到观测到的仙女座星系蓝移量的一半。但是星系确实在空间中运动，这些运动产生的多普勒效应叠加在宇宙红移上了。

除了离我们最近的邻居，宇宙学红移比任何多普勒效应都要大得多，并占了主导地位。但仙女座星系的多普勒效应比宇宙红移要大得多。仙女座星系的确正在快速地向我们移动，并将在大约 40 亿年后与银河系相撞——巧合的是，那时的太阳也正接近其生命的尽头。这样的碰撞，发生在大致势均力敌的圆盘星系之间将导致二者合并。如果每个星系中，恒星之间的距离都像银河系中那样大的话，那么合并时，两个星系中的恒星基本不会发生碰撞。但是，计算机模拟显示，当两个星系合并成一个系统时，引力会导致两个星系的盘状结构被彻底破坏，最终形成一个巨大的椭圆星系。

41

即便仅从人类家园银河系的角度出发，本章所描述的发现也称得上非常重要。原因在于，有坚实的证据表明，银河系是一个非常普通的圆盘星系，代表一类典型的圆盘星系。正是因为如此，我们可以在近距离观察的基础上，自信地利用对自己星系结构和演化的知识，来全面了解圆盘星系的起源和性质。我们在宇宙中毫不起眼且并不特殊，虽然这种观点直到 20 世纪末才得以确立。

04

小插曲：平庸的银河系

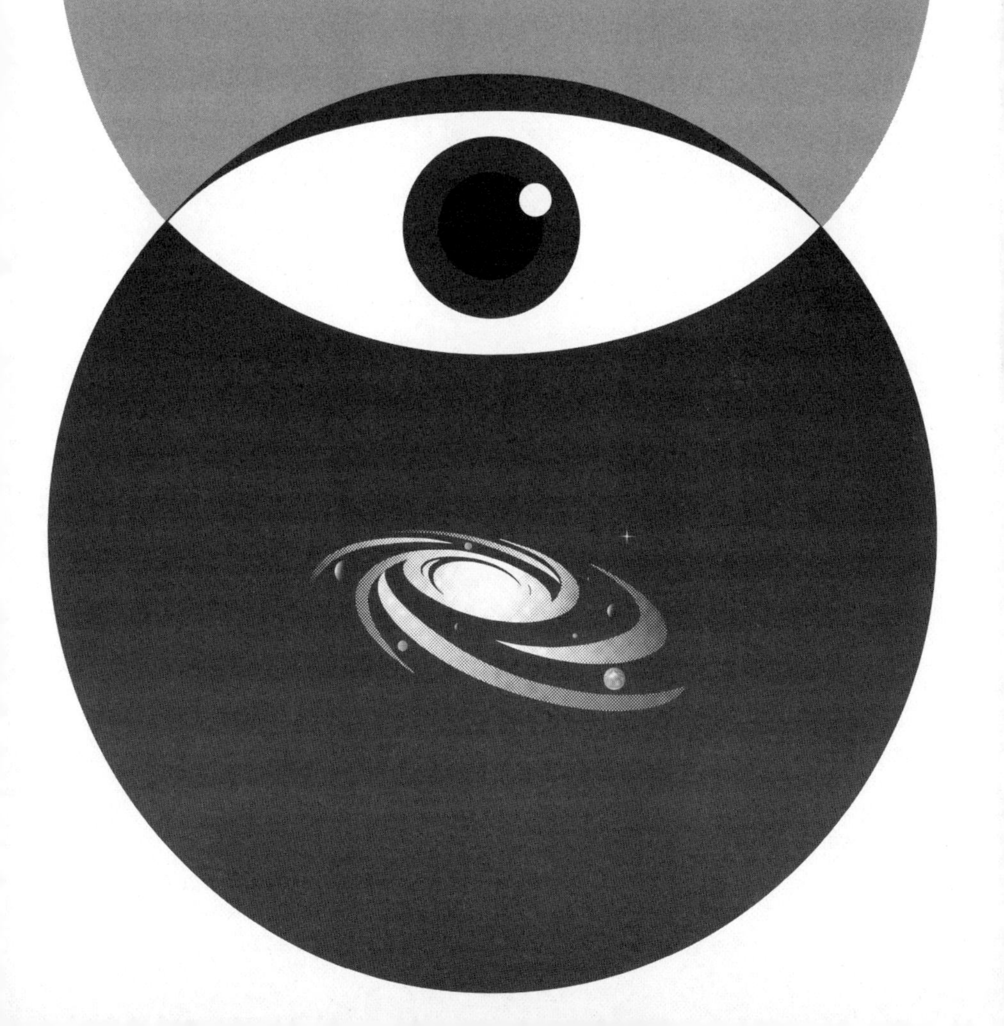

42

可以说，科学革命始于1643年。那一年尼古拉斯·哥白尼（Nicolaus Copernicus）出版了一本名为《天体运行论》（*De Revolutionibus Orbium Coelestium*）的书，其中列出了地球不在宇宙中心，并且围绕太阳运行的证据。从那以后，人们逐渐意识到：太阳只是一颗普通的恒星，在银河系中都没有什么特殊的地位，更不用说在浩瀚的宇宙中了。人类也只是地球上的生命之一，以往的认知，不过是我们狭隘的偏见。有些天文学家讥诮地说，所有这些都不过是支持"地球平庸原则"的证据而已。该原则认为，就宇宙而言，我们周围的环境完全没有任何特殊性。对那些仍固守地球中心思想的人来说，平庸原则无疑是一种耻辱。但如果平庸原则是正确的，我们的确可以借此从对周围环境的观察中，推断出关于宇宙整体性质的有意义的结论。如果银河系是平庸的，那么其他数以亿计的星系肯定与银河系非常相似，这就像城市的郊区看起来都非常相似一样。

但在哈勃测算出宇宙距离标度后长达几十年的时间里，银河系似乎仍然享有特殊地位。这是因为，

[43] 用哈勃的标度值来衡量，其他星系离银河系的计算距离比实际距离近。据此得出的其他星系的大小都比实际要小，这使得银河系看上去似乎是宇宙中最大的星系。现在我们都知道哈勃错了。由于面临重重困难——包括恒星消光以及造父变星与其他变星的严重混淆等，哈勃最初得到的哈勃常数值，比今天已被普遍接受的数值大了7倍左右。换句话说，哈勃测算出的所有银河外天体的距离都只有实际距离的1/7。这怪不得哈勃，这本来就不是一蹴而就的事。几十年来，随着观测手段的进步以及一个又一个纠错，宇宙距离标度经历了一个缓慢修正的过程。我并不打算把整个过程都罗列一遍，而是想用最新、最精确的观测资料，为银河系的平庸提供最简单、最直接的证据。

其实早在20世纪30年代，就有一些科学家对银河系可能是一个超级大星系的想法不以为然。其中反应最激烈，并提出最有力质疑的人是天文学家亚瑟·爱丁顿（Arthur Eddington）。他曾在1919年率领日食远征队，证实了爱因斯坦广义相对论的预言，这是爱丁顿最广为人知的壮举。爱丁顿坚信地球平庸原则，并在1933年出版的《膨胀的宇宙》（*The Expanding Universe*）一书中写道：

来自天文学方面的教训一再提醒我们要谦卑，以至于不知不觉间我们接受了这样的观点，那就是银河系并不拥有特殊地位，至少大自然在计划中并未把银河系放于数以百万计的其他星系之上。但天文观测似乎难以证实这一点。据目前的观测，旋涡星云虽然与银河系大体相似，但明显要小一些。有人说，如果旋涡星云是岛屿，我们所在的星系就是大陆。我想，我们至多算中产阶级，因为我相当不喜欢我们是宇宙贵族这种说法。地球是行星中的中产阶级，既不像木星那么巨大，也不像小行星那样渺小；太阳是一颗中等大小的恒星，虽不像五车二[1]（Capella）那样庞大，但也谈不上渺小。因此，那种我们恰巧居住在一个非凡星系中的观点似乎是错误的。坦白地说，我一点儿都不相信，这也太巧了。我认为，银河系与其他星系的关系是极具研究价值的问题，对其进一步观测研究，我们将能发现更多线索。最终我们将发现，宇宙中有许多大小与银河系相当或比其更巨大的星系。

44

1　指御夫座 α。其英文名称源自拉丁文，原意是小山羊。

爱丁顿的观点很有道理，它最终被证明是对的。但在 1933 年，他还只是提出了“中产阶级”说。毕竟，一些星系比其他星系更大。你当然可以争辩说，如果宇宙中真的存在一个无与伦比的超大星系统治者，由一大群小星系环侍周围，我们将更有可能发现自己在大陆上，而不是在其中的一个小岛上。平息这一争论的唯一方法，就是将其他圆盘星系的大小与银河系进行对比，参与对比的星系数量越多越好。要做到这一点，就需要对圆盘星系进行精确的距离测量，

图 8　沿太空轨道绕地球运行的哈勃太空望远镜

这可以归结到造父变星的距离测量问题。这意味着，1990 年发射哈勃望远镜，或在 1993 年它还未被修复之前，我们无法测量造父变星的距离。

在哈勃开拓性的工作完成了半个多世纪之后，精确测定宇宙距离标度的需求仍然十分迫切，这正是建造哈勃太空望远镜的主要原因。建造哈勃太空望远镜的明确目标之一，是获取来自至少 20 个星系中的造父变星数据，并利用这些数据将哈勃常数提高到 ±10% 的精度。等到这一重要项目的观测阶段结束时，哈勃研究小组已经利用造父变星精确地确定了 24 个星系的距离。下一阶段，是利用这些数据校准超新星等其他示距天体的标度。这期间，造父变星的基本数据已经实现了共享。1996 年，我和苏塞克斯大学的西蒙·古德温（Simon Goodwin）以及马丁·亨德利（Martin Hendry）一起，使用这些造父变星的距离数据，响应爱丁顿所倡导的“进一步观测研究”，验证他的看法，即银河系只是一个普通的旋涡星系。（研究结果已于 1998 年发表）

45

研究中使用的主要是来自哈勃太空望远镜的数据，同时结合了来自地面望远镜的部分数据。我们选择了 17 个外观很像银河系的旋涡星云，而且都已有了较精确的距离数据。实际上，测量星系角直径的标准方法是在其周围绘制亮度等高线（等照度），并以某一特定的亮度等高线为边界，进行角直径测量。有了用这种方法确定的角直径，以及利用造父变星测得的精确距离之后，这17个星系的实际大小也就随之确定了。

该项目最困难的部分是测量银河系的等值直径。困扰我们的问题一直是同一个——我们无法看到银河系的全貌，正所谓“不识庐山真面目，只缘身在此山中”。不过，通过综合分析银河系内恒星分布的观测数据，可以推算出俯视它的样子。结果显示，银河系的等照度等值直径略小于27千秒差距。接下来面临的重大问题就是：如何将银河系的直径与样本中其他17个星系的直径相比较？简而言之，就是要将银河系的直径与平均值进行比较。包括银河系在内，样本中所有18个星系的平均直径，刚好超过28千秒差距。正如爱丁顿所猜测的那样，银河系只是一个普通的旋涡星系，其直径略小于平均值，但差异并不显著。可以肯定的是，银河系不是被岛屿环绕的大陆，但其大小也没有明显低于平均水平。总之，银河系就是普通的星系。

除此之外，这些星系直径的观测结果，还可以用来确定哈勃常数的值，而且精确度能在哈勃重点项目

46

设定的 10% 的目标之内。在下一章中，我会将哈勃常数置于宇宙学的背景下，从而揭示宇宙本身的年龄，即大爆炸以来已经逝去的时间。

05

膨胀的宇宙

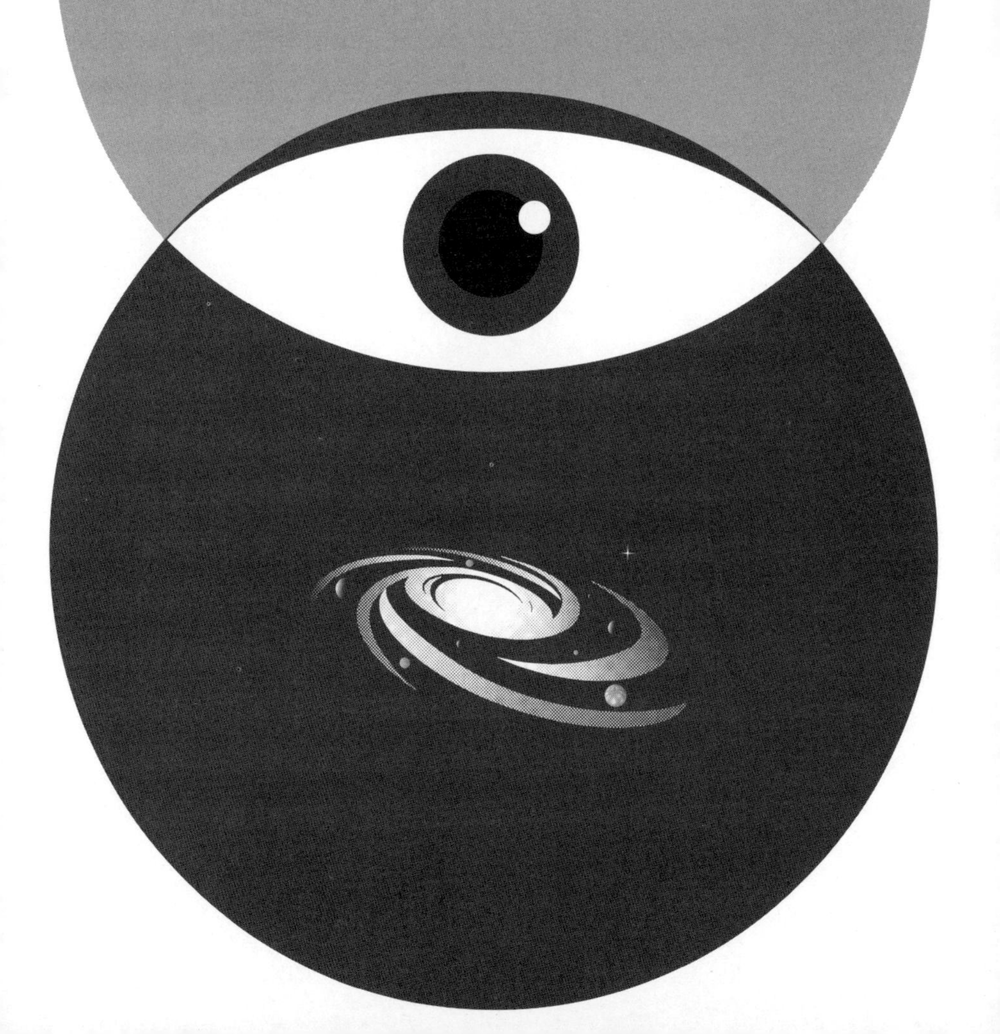

现代宇宙学发端于哈勃关于星系的两个重大发现：第一，星系是太空中独立于银河系之外的岛屿；第二，来自遥远星系的光的红移量，与这个星系和我们之间的距离存在某种关系。将这两个发现放在一起考虑，并把遥远的星系当作测试粒子（test particle），就可以揭示宇宙的整体行为。特别是，这样做还可以看出宇宙正在膨胀。

47

红移－距离关系在20世纪20年代末被发现。当时的人们对此感到惊异之余，也立刻意识到，描述这种普遍行为的数学理论已经产生了，这就是广义相对论。爱因斯坦的广义相对论是描述空间、时间、物质和引力关系的普遍理论。这一理论的关键特征在于，空间和时间不再彼此独立，而是结合在一起，成为不可分割的、称为时空的四维实体。四维时空的概念可以追溯到1908年，当时，赫尔曼·闵可夫斯基（Hermann Minkowski）完善了爱因斯坦于1905年发表的狭义相对论。“从今往后，”闵可夫斯基说，“孤立的空间和独立的时间注定要黯然失色，只有二者相互交织组合的统一体才能闪耀于物理现实之中。”

48 狭义相对论的局限性，在于它不能处理包含引力或有加速度时的问题。这即是它“狭义”的原因，狭义相对论只在惯性系中有效，处理的是平直时空中的问题。在这个范围内，我们将所有电磁辐射统称为光。狭义相对论精确地描述了所有运动物体与光之间的关系——当然，这些物体必须以恒定的速度在直线上运动。另外，利用狭义相对论，也可以从运动物体的角度观察世界。狭义相对论的伟大成就，远非这几句话就能概括。在这个理论中，爱因斯坦从本质上修正了艾萨克·牛顿（Isaac Newton）的动力学，并将詹姆斯·克拉克·麦克斯韦（James Clerk Maxwell）对光的理解考虑了进去。不过狭义相对论只是一个过渡理论，更完备的理论还需要将引力和加速度纳入其中。

经过多年的不懈努力，爱因斯坦在1915年成功地将引力和加速度纳入了新的理论。这就是著名的广义相对论。理解这一理论最简单的方法，就是借助闵可夫斯基的四维时空。在广义相对论中，这种时空是有弹性的，会被存在的物质扭曲。在这种时空中，物体会沿着弯曲的路径运动，就像弹珠在蹦床上滚动时，会自动沿着沉重保龄球压痕上的弯曲路径运动一样。通常所说的引力，在广义相对论中其实是时空弯曲造成的一种效应。用一句名言来概括广义相对论就是：

“物质告诉时空如何弯曲，时空告诉物质如何运动。”[1]

最为关键的是，在被物质扭曲的时空中，光线也会沿着弯曲的路径运行。然而这种效应非常微弱，实际上很难观察到，除非出现以下三种情况：其一，所涉及的物质数量非常庞大；其二，物质虽然体积较小但密度异常高；其三，前两者同时具备。在太阳系里，这种扭曲也只有在太阳附近才能勉强观察到。广义相对论预言，由于太阳的质量扭曲了其附近的时空（图 9），从太阳旁边经过的星光会发生一定程度的偏折。因此从地球上看恒星时，其太空背景上的表观位置会在太阳经过时发生移动。当然，由于阳光太过耀眼，那些恒星的移动是很难被看到的。实际观察这种移动的唯一方法，就是在日全食期间，当太阳的光线被月球挡住时进行观察。幸运的是，在 1919 年恰好发生了一次日全食。当时，由亚瑟·爱丁顿率领的一支远征队观测到了这种效应，结果与爱因斯坦理论的预测完全吻合。从那时起，爱因斯坦就成了万众瞩目的名人——尽管很多人并不知道他为什么出名。也正是从那时开始，广义相对论通过了各种各样精心设计的实验的检验。最

49

1 相对论大师惠勒（John Archibald Wheeler）曾用这句精辟的话来总结爱因斯坦的场方程式。

近一次是在太空中进行的，验证的是广义相对论关于地球引力对失重陀螺仪影响的预测。[1]

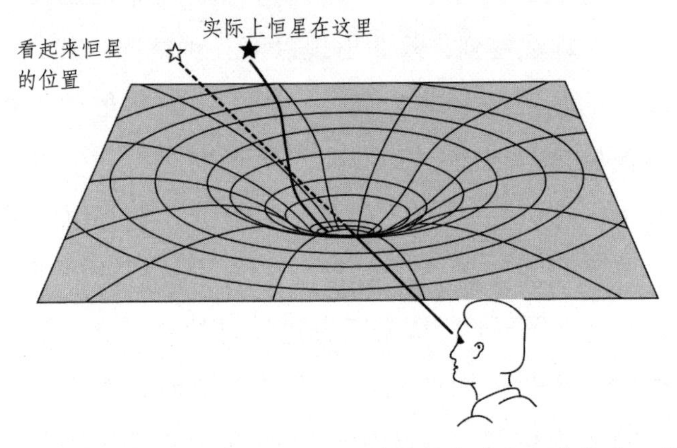

图 9　太阳对附近时空的扭曲，就像重物在蹦床上造成的凹痕，使得来自遥远恒星的光在太阳附近沿着曲线运动。或者说，从地球角度看，当太阳从恒星旁边经过时，恒星的位置会发生改变

广义相对论是描述空间、时间和物质整体行为的最佳理论。正如爱因斯坦所要求的那样，该理论自然地给出了对宇宙的描述。这里所说的宇宙是指所有空间、时间和物质的总和。麻烦的是，广义相对论描述的宇宙有很多种。爱因斯坦的广义相对论方程有许多解，这在数学中是常有的事。比如众所周知的例

50

1　指 2004 年美国发射的一颗“重力探测 B 型”的卫星，用以检测引力对失重陀螺仪的影响。

子，方程 $x^2 = 4$ 有两个解，$x = 2$ 和 $x = -2$，这是因为 2×2 和（−2）×（−2）都等于 4。爱因斯坦的方程远比这复杂，解也更多。其中一些解描述了膨胀的宇宙，另一些则描述了收缩的宇宙，还有一些描述了在膨胀和收缩之间振荡的宇宙，林林总总。但是，爱因斯坦失望地发现，没有一种解描述的宇宙是静态的。

这个结果令他非常困惑，因为爱因斯坦 1917 年得出这些解的时候，几乎所有人都认为宇宙应该是静态的。大多数天文学家仍然坚持，银河系就是宇宙的全部。尽管其中的恒星都在不停地运动，但从总体上看，银河系既没膨胀，也没收缩。为了将宇宙的静态解纳入广义相对论的框架内，爱因斯坦人为地在方程中加入了一个额外项。现在这一项被称为**宇宙常数**（cosmological constant），通常用希腊字母 Λ[1] 表示。12 年后，哈勃发现了红移－距离关系，用事实证明，不包含宇宙常数 Λ 项的广义相对论场方程最简单的解之一，与膨胀宇宙的数学描述相吻合。因此，爱因斯坦将宇宙常数的引入描述为他职业生涯中“最大的错误”。那一时期，除了少数因兴趣爱好还保留这个常数的数学家之外，差不多所有人都抛弃了宇宙常

1 读作“拉姆达”。

数——不管它是否描述了真实的宇宙。

广义相对论为宇宙提供了很好的描述，这一发现的全部含义几乎都被写在了科尔斯的书中。然 51 而，关键在于，方程所描述的膨胀是空间随时间的膨胀。宇宙学上的红移，并不是由星系在空间中运动引起的多普勒效应。不要将星系的退行想象成爆炸中四处飞散的碎片，而要将其想象成是空间本身的膨胀裹挟着星系远去。红移的成因应该描述为：当光在空间中传播时，因为空间本身不断膨胀，使得光的波长被拉长，在光谱上移向红端。

然而，这种空间对光的拉伸与红移之间的关系，与相对论效应有关。如果把红移转化为等效速度，那么只要等效速度与光速相比很小，就可以忽略相对论效应。红移通常用字母 z 表示。如果 $z = 0.1$，意味着物体的退行速度是光速的 1 / 10，即大约 30 000 千米 / 秒，比哈勃和哈马逊在他们开创性的研究中测量到的任何速度都要大。0.2 的红移代表物体正在以两倍于 $z = 0.1$ 时的速度退行，依此类推，直到某个极限为止。既然没有什么东西能比光的速度更快，那么如果前面 52 所说的简单规则成立的话，可能产生的最大红移应该就是 1。但事实并非如此，考虑相对论效应时，与以光速退行相对应的最大红移是无穷大。事实上，一旦等效速度超过光速的 1 / 3，相对论效应就变得无法忽略了。当相对论效应显现出来时，情况就不同了。例如，数值为 2 的红移，并不是说物体以两倍光速的速度退行，而是以光速的 80% 退行；数值为 4 的红移对应的退行速度为光速的 92%。目前已知的最大红移量比 10 还大，这极其罕见。

事实上，宇宙中几乎没有孤立的星系，大多数星系都以星系团的形式出现。每个星系团可能包含几个到数千个星系，这些星系通过引力聚集在一起。星系团内的单个星系围绕着共同的质心运动，而整个星系团被空间的膨胀所裹挟。就像蜜蜂群，单个星系彼此绕行，而整个群体作为一个整体移动。星系团中星系发出的光，整体上有一些平均红移，但这是宇宙膨胀引起的宇宙学红移。不过有些星系的红移稍微大一些，有些星系的红移稍微小一些。其中，向我们奔来的那些星系红移较小，原因是它们在空间中的运动产生了多普勒蓝移，从而抵消了一部分红移。远离我们而去的那些星系红移较大，是因为它们在空间中的运动产生了多普勒红移，从而加大了总红移。当天文学家使用“红移与星系的距离成正比”这个简化说法时，实际上已经把所有这些影响因素都考虑在内了。

关于宇宙膨胀的另一个关键点是：膨胀没有中

53 心。星系退行的红移量与离银河系的距离成正比。而且，无论身处哪个星系，你都会看到同样的结果——红移与距离成正比。这一事实也再一次展示了“地球平庸原则”。这一点可以用简单的类比说明。想象在一个球体的表面，随机地画上一些不同颜色的点来表示不同星系。如果让球体膨胀起来，每个点之间的距离都会增加，这与膨胀宇宙中星系的行为非常相似。假设膨胀使每个点之间的距离加倍，相距 2 厘米的点最终会相距 4 厘米，相距 4 厘米的点最终会相距 8 厘米，依此类推。如果在膨胀之前，有 3 个点以 2 厘米的距离排成一条直线。那么膨胀之后，中心点到两个端点之间的距离都是 4 厘米，那么两个端点之间的距离将是 8 厘米。从一个端点观看，中心点退行 2 厘米时，另一个端点将退行 4 厘米（图 10）。两端点之间的距离是到中心点的两倍，相对于中心点来说，“红移”量也是两倍。从球体表面的任何点观察，整体情况是相同的，红移与距离成正比。

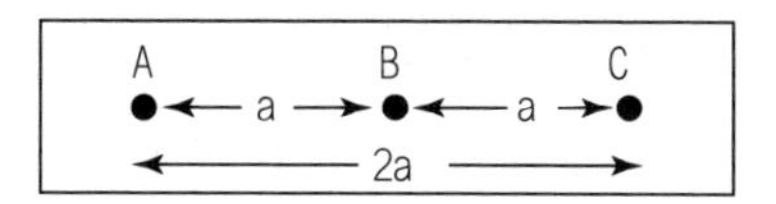

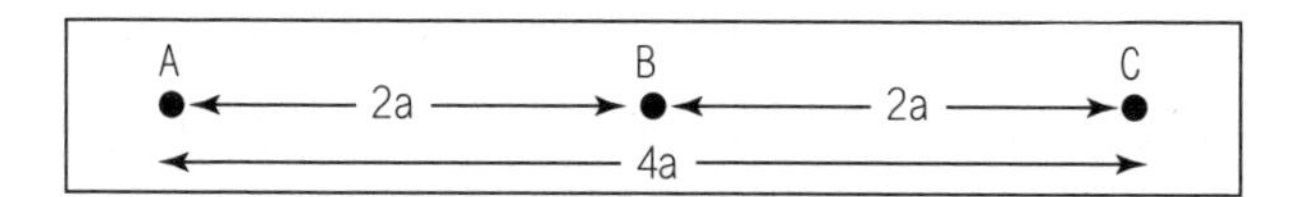

图 10　时空膨胀的样子就如同拉伸一块橡胶，星系 A、B、C 并不随着其拉伸在这之间的空间移动。当空间扩大到使 A 和 B 之间距离变为两倍时，也会使包括 A 和 C 在内的其他两个星系之间的距离翻一番。从宇宙中每个星系的角度来看，其他星系都在以与其距离成正比的速度退行。例如，因为 C 离 A 的距离是离 B 的两倍，所以，当所有的距离加倍时（即标度因子加倍时），C 从 A“移开”的速度似乎是从“B”移开的速度的两倍

但是，如果将上述过程倒转一下，缩小球体的大小，会怎样呢？显然这些点会逐渐靠近，蓝移与距离成正比。这就相当于倒转宇宙膨胀的历史。很明显，如果星系现在正在相互分离，那么它们过去一定靠得更近。虽然不是那么直观，但如果逆着时空膨胀的方向从现在向过去回溯，依据广义相对论，就必定会到达一个时空点。在这个时空点处，所有的物质和所有的空间及时间都收缩成一个体积无穷小、密度无限大的理论意义上的奇点，就与之前所提到过的藏在黑洞中心的奇点一样。由于物理学家不相信那些导致无穷大之类极端物理推论出现的理论，所以，通常认为广

54 义相对论在这种情况下肯定失效了。不过，我们有充分的理由相信，宇宙诞生于一个体积极小（比原子还小）、温度极高、密度极大（包含了今天宇宙中所有的质量）的奇点——尽管这些物理量中没有一个是允许无限的。认为宇宙有超密、超热开端的想法是大爆炸模型的核心。大爆炸构想是在20世纪下半叶开始被重视的，大量的观测数据证实了宇宙膨胀的事实。宇宙学家需要努力回答的重大问题包括大爆炸是什么时候发生的，宇宙年龄有多大等。这些问题的答案来自对星系的研究，这些研究也提供了对哈勃常数的测量方法。

哈勃常数（H）是衡量宇宙膨胀速度的一个指标。如果宇宙一直以相同的速度膨胀，或者说哈勃常数从未变化，那么取哈勃常数的倒数（1/H），就可得到大爆炸之后经过的时间，也就是宇宙的年龄。类似地，假设一辆汽车以60英里/小时（约97千米/小时）的速度沿着M4高速公路离开伦敦向西行驶，当汽车驶出伦敦120英里（约193千米）时，很容易就能算出这段旅程是在大约两小时前开始的。但宇宙年龄的估算稍微复杂一点儿。因为从广义相对论场方程中得到的最简单的宇宙模型表明，早期的宇宙一定膨胀得更快，随着时间的推移，由于引力阻碍了膨胀，膨胀速度会逐渐减慢。对宇宙年龄更好的估计是1/H的2/3，而1/H本身叫作哈勃时间。问题的关键在于，只要能测量哈勃常数，就能测量宇宙的年龄。

因为宇宙的年龄与哈勃常数成反比，哈勃常数的值越小，宇宙就越老。根据哈勃自己确定的值，即525千米/（秒·百万秒差距），宇宙的年龄大约是20亿年。即使是在20世纪30年代，这个估计值也有很明显的问题，因为宇宙年龄不可能比地球还小。为什么大爆炸的概念直到20世纪40年代末才开始受到重视，这是其中一个原因。在消除了各种变量导致的误差之后，距离标度得到了大幅度的修正。哈勃常数一下子减半，宇宙的估计年龄翻了一番，使宇宙看起来和地球一样古老。

55

大约在同一时期，天文学家已经能够很好地理解恒星是如何运行的，并有能力对恒星的年龄做出可靠的估计。20世纪50年代，有人偶然发现，一些恒星的年龄竟然超过了100亿年，这再次使“大爆炸”构想陷入了尴尬境地。因此，当时宇宙学的稳态模型成功吸引了一大批天文学家。稳态模型的逻辑是，当星系在不断膨胀的宇宙中相互远离时，负责空间拉伸的力也会使得新物质在星系之间的空隙中产生，形成由氢原子构成的气体云，而新星系将从气体云中形成以填补空隙。在这个模型中，宇宙没有开始，也没有结束，从整体上看是不变的。这个稳态模型的丧钟在20世纪60年代敲响，当时有无线电天文学家发现了来自太空各个方向的微弱的无线电噪声，这就是**宇宙微波背景辐射**。大爆炸理论曾预测了这种辐射的存在，尽管当时这个预言已经被遗忘。从理论上讲，这种辐射正是大爆炸时能量辐射被红移后的残留物。此后的观测数据不断证实了这个理论，其中就包括被送入太空，专门用来研究宇宙微波背景辐射的卫星提供的珍贵数据。由于宇宙年龄的估计值随着时间的推移而逐渐增加，人们对稳态模型的认同度也就逐渐降低了。

1950年左右，在精度不断提高的观测数据的基础上，人们开始逐步修正距离标度，将哈勃常数的值不断拉低。直到20世纪90年代初，人们才知道，使用常用单位，哈勃常数的值应该介于50～100之间。正如一位天文学家所指出的，75±25，这就是哈勃重点项目的关键值。

56

比如仙女座星系，星系团中的星系通常在太空中以几百千米/小时的速度随机运动。这意味着，为了得到星系团的宇宙学红移的可靠估算值，研究遥远的星系团是最佳选择。这样得到的宇宙学红移值较大，而单个星系的随机速度及其相关的多普勒红移在整体红移中所占的比例较小。然而，要测量更远处的星系团的距离更加困难。所以用星系团来测定哈勃常数的值时，就需要再三权衡。哈勃重点项目运用的还是哈勃设计的传统技术，利用造父变星测定邻近星系的精确距离，再用这个距离来校准其他示距天体的亮度，比如超新星。人们就这样通过一系列步骤逐步深入地了解宇宙。不同之处在于，在哈勃之后的 60 年里，研究人员拥有了更好的望远镜，不同类型的变星之间的混淆问题也得到了解决。他们懂得了恒星消光，对超新星等次级示距天体的理解也比哈勃时代要好得多。2001 年 5 月，重点项目团队提出的哈勃常数的最终估算值为 72 ± 8，对应的宇宙年龄约为 140 亿年。令人高兴的是，在 20 世纪 90 年代，由完全不同的技术测得的最古老恒星的年龄，大约是 130 亿年。这在一定程度上佐证了哈勃常数的测量值。宇宙确实比它所包含的恒星和星系更古老。

这一结果的影响远比人们一开始预计的要深远得多。宇宙的年龄是通过研究宇宙中一些最大的天体——星系团，并利用广义相对论分析它们的行为判断的。而通过研究宇宙中最小的物体——原子的原子核，来理解恒星的运行方式，就能在此基础上计算出恒星的年龄。使用20世纪另外一个伟大的物理学理论——量子力学，则可以计算原子核是如何融合并释放出巨大的能量，从而维持恒星持续发光的。宇宙的年龄和恒星的年龄非常协调，且最古老恒星的年龄比宇宙的年龄略小，这一事实是肯定整个20世纪物理学工作的最令人信服的理由之一。物理学对世界的描述，从最小的尺度到最大的尺度，都已取得了突破性进展，改变了人们对世界的认识。

58

现在，确定哈勃常数数值接近70千米/（秒·百万秒差距），这个结论已被其他技术所证实。这些技术既涉及高科技设备卫星，也涉及人们对物理学的深刻理解。但是，有一种简单的方法强调了星系和宇宙之间的关系，并且当其与更复杂的测量技术相结合时，也能证实我们星系的平庸。

银河系只是一个普通旋涡星系，用宇宙学术语来说，证据是来自邻近星系的相当小的样本。我们姑且接受这一结论，以便得到一种估算银河系到其他星系距离的方法，即将遥远星系的大小与银河系的大小比较，或者与我们本地样本（邻近星系）的平均大小相比较，这两个比较的结果非常接近。与单个星系进行这样的比较是没有意义的，因为星系的大小差别很大。在我们的宇宙邻居中，旋涡星系 M101 最大，直径接近 62 千秒差距，是银河系的两倍多。所以假设 M101 和银河系一样大，并以此估算两者之间的距离，明显不是个好主意。天文学家们需要的是某种统计测量法，为的是将遥远星系的平均大小与邻近星系的平均大小进行比较。

图 11　不规则星系 NGC 1427

自哈勃时代以来，观测者们建立了一系列观测目录，收集了数以千计的星系的位置、红移和角直径大小等信息。这里所说的角直径通常用等光直径来表示。这样的目录很多，每个都包含了数千个星系。其中一些目录中的角直径可以作为银河系平庸的证据。角直径乘以一个数字，就可以转化为线性直径。这个数字只依赖于已知的红移量和假设未知的哈勃常数。如果取散布在天空中的数千个有着不同红移量的星系，就可以选择哈勃常数的某个值，计算出所有星系的线性直径，然后在整个样本中取 59 平均值来估算星系的平均大小。用计算机来做这项工作是很简单的。计算时，只需不断变换哈勃常数值，一次又一次地重复计算，直到计算出来的平均值等于类似银河系这样的旋涡星系的平均直径即可。这样就给出了哈勃常数的唯一值。

这里也有些必须克服的实际困难。首先，必须确保所有角直径的测量方法一致；其次，纳入样本中的 60 星系，必须与本地样本中的星系具有相同的整体结构；最后，样本中的观测资料需要囊括所有相关星系。最后一点最难做到，其中一个最重要的原因是大星系比较容易看到，而一些红移较大的小星系本来也是该进入样本的，却没有囊括进来。这是因为红移较大的星系更遥远、更暗淡，也更容易被忽视。这种效应就是所谓的**马尔姆奎斯特偏差**（Malmquist bias）。

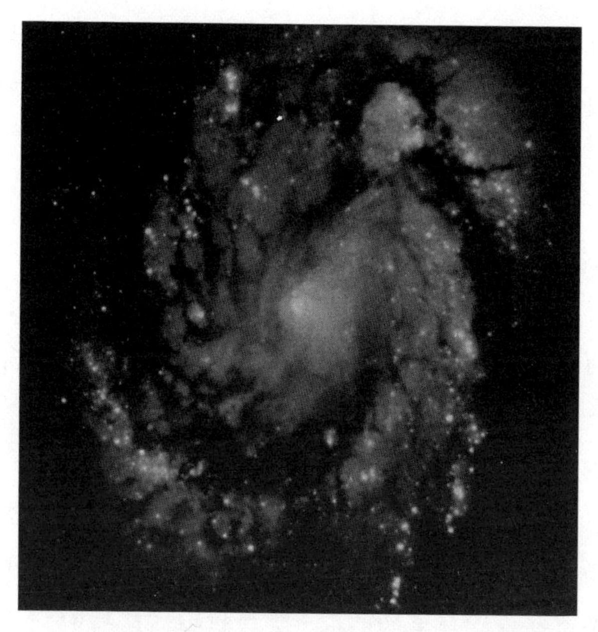

图 12 星系 M100 的中心区域，由哈勃太空望远镜上的 WFPC2 成像

幸运的是，通过比较不同红移量的大大小小星系的数量，可以计算出这种效应的统计数据，得出小星系伴随红移量的增加而从样本中减少的规律，并据此对结果加以修正。麻烦之处在于，这项技术仅适用于计算与我们相隔大约 100 兆秒差距的星系，附近的星系必须被排除在计算之外，因为它们的随机多普勒频移与宇宙学红移大小相当，会造成数据混淆。即使有这些条件限制，一份标准目录 RC3 依然收录了一个远超 1000 个星系的子目录，其中每个星系都满足上述所有

条件。这对统计上可靠的样本来说足够了。当一切准备就绪后，在20世纪60年代中期，人们得出了基于星系直径比较的哈勃常数值，这个值与其他测量值一致。当然，得出这一结果的前提是：银河系确实只是一个普通旋涡星系。

这并不是测量哈勃常数最佳或最精确的方法，但却有自己的价值，原因有两个。首先，这是一种很好的、可以用日常经验来理解的物理技术。就像站在原野尽头的一头奶牛看上去很小，只是因为它离得太远了，不需要高深的物理或数学知识就能理解。其次，这个论点可以颠倒过来。第一个真正证明银河系只是一个普通旋涡星系的证据，来自将银河系与17个相对较近星系进行的比较。但如果哈勃常数接近70，那么正如精密观测和深入分析所表明的，可以利用该值来计算样本中1000多个星系的平均大小（其中一些远在100兆秒差距之外），其结果也的确非常接近银河系的大小和附近样本的平均大小。至少，银河系是典型的圆盘星系，位于直径约为200兆秒差距、体积超过400万立方秒差距的太空的“局部”区域。

61

与可观测宇宙的大小相比，这确实只是一个局部区域。有一些已知的天体，其实测红移与大于100亿光年的距离相对应，比用来估算哈勃常数的最远星系要远 30 倍。对这些天体的研究表明，事情还远不止于此。从大爆炸以来，宇宙的膨胀似乎并没有像爱因斯坦方程的最简单的解所预测的那样放缓，相反，它可能已经开始加速了。

20 世纪 90 年代，天文学家开始利用超新星的观测数据校准红移大约为 1 的红移 - 距离关系，这类超新星已知的最大红移都小于 2。这项技术依赖于一种特定超新星的发现——一个名为 SN1a 的超新星家族。表面看去它们的绝对亮度峰值都相同。这类超新星都是借助近邻星系 SN1a 被发现的，而星系 SN1a 的距离我们很清楚。这一发现尤其重要，因为这种超新星非常明亮，在极其遥远的地方都能看到。

因为所有 SN1a 超新星都有相同的绝对亮度，所以它越是暗淡，离我们就必然越远。这其实是说，只需找出某星系中的 SN1a 超新星，那么就可以根据其表观亮度峰值计算出其所在星系的距离。如果也能同时测量同一星系的红移，就能校准哈勃常数。尽管穷尽了所有技术手段，但观测数据依然表明，如果星系位于哈勃常数公认值所导出的距离上，那么遥远星系中超新星的亮度，还是要比预期值暗淡一些。

62

不能排除这样一种可能性，即遥远星系中的超新星确实不如邻近星系中的超新星明亮。但最符合所有现有证据的结论是，如果宇宙自大爆炸以来一直按照最简单的宇宙学模型膨胀，这些超新星就应该离我们更远一点儿。只需要对爱因斯坦方程进行些许修改，就能将一切拟合——必须把当初那个不起眼的宇宙常数重新放回到方程中去。也许这根本就不是一个错误。

爱因斯坦在广义相对论场方程中引入宇宙常数，是为了适应静态宇宙模型。但是常数值不同，宇宙膨胀的快慢也不同，甚至在有些常数值下宇宙会

63

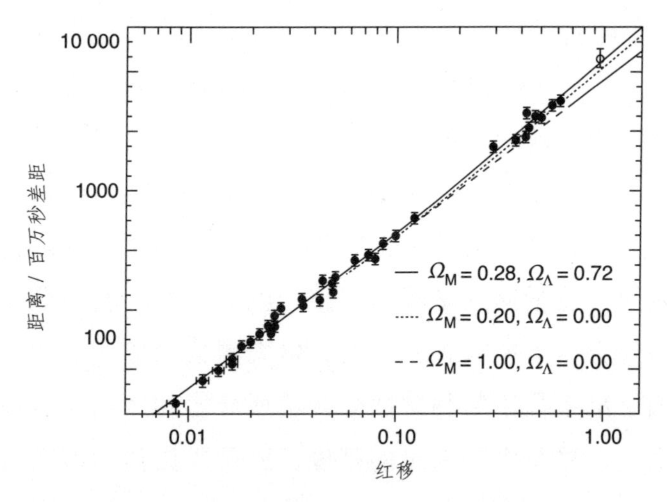

图 13　利用对高红移超新星的观测，红移–距离图可以扩展到宇宙中极其遥远的地方。对数据（实线）的最佳拟合考虑了宇宙学常数 Λ 的影响

收缩。为解释超新星观测数据而添加进方程中的宇宙常数，则暗示整个宇宙里充满了某种能量，也就是一种对普通物质没有明显的影响，但能与万有引力相对抗，把宇宙向外推，就像被压缩的弹性流体那样。因为宇宙常数传统上用希腊字母 Λ 表示，所以也称为拉姆达（Λ）场。只要为这个场选择合适的密度，就可以简明地解释宇宙在大爆炸后最初几十亿年的膨胀是如何放缓——就像简单模型所预测的那样。但随后，Λ 开始让膨胀非常缓慢地加速。

其工作原理是这样的（关于宇宙加速还有更复杂的解释，但由于最简单的解释已经很好，所以我不打算在这里讨论那些复杂的）：拉姆达场是一个常量，自大爆炸以来始终保持相同的值。这个场是不可见的，通常称之为“暗能量”。暗能量是时空本身的一种固有属性，也就是说，相同“大小”的时空蕴含相同数量的暗能量，与其他因素无关。所以当空间膨胀产生出更多空间时，暗能量并不会被稀释。这就是说：第一，储存在单位体积空间中的暗能量永远保持不变；第二，每立方米空间中的暗能量，总是向外施加相同的力。这与宇宙膨胀时物质所经历的变化截然不同。当宇宙从大爆炸中诞生时，所有物质的密度都像原子核的密度那样大。一点儿这样的物质所包含的质量就相当于今天地球上所有人的质量和。因此这种超高密度物质的引力完全淹没了拉姆达场。随着时间的推移，宇宙膨胀了，同样数量的物质占据的空间越来越大，物质的密度随之下降。由此，引力对膨胀的影响逐渐变小，直到小于暗能量的影响。

64

超新星的观测数据表明，在大约五六十亿年前，物质引力对膨胀的减缓程度和暗能量对膨胀的加速作用相同。用红移领域的术语来说，物质引力与暗能量作用的主次转换发生在红移量 0.7 左右。从那以后，暗能量的影响一直大于物质引力的影响，使得宇宙的膨胀加速。

如果宇宙真的在加速膨胀，那就说明，宇宙的年龄比没有考虑加速时计算的 140 亿年稍微老了一点儿。因为，如果宇宙过去膨胀得更慢，就需要更长的时间才能达到现在的状态。但是这种影响非常小，从宇宙的年龄比最古老的恒星的年龄大来看，理论研究的大方向是正确的，所以我们大可不必担心。

使宇宙加速膨胀所需的暗能量非常少。考虑到爱因斯坦的质能关系，能量和质量是等价的，与暗能量相关的质量在宇宙的每立方厘米空间中小于 10^{-29} 克，也就是说，每立方厘米中只有不到 0.000 000 000 000 000 000 000 000 000 01 克。所以不足以使地球、太阳系、银河系，甚至是星系团这类物质膨胀和分裂，因为在局部尺度上，聚集在一起的物质的引力完全压倒了暗能量。

然而，从宇宙尺度上看，尽管单位空间中的暗能量形式的质量如此之小，但在宇宙中恒星与恒星、星系与星系之间“空”的空间实在太大了，其中的暗能量形式的质量的总和，远远大于以明亮的恒星和星系形式存在的物质总和。这会让哈勃和他同时代的人大吃一惊，想必他们误以为自己研究的才是物质宇宙中最重要的组成部分。但是直到 20 世纪 90 年代末，人们才清楚地认识到，宇宙中有比可见物质更多的东西，宇宙学家开始努力寻找所谓的“缺失的质量”。事实证明，拉姆达场就是现代宇宙图景中缺失的部分，它为理解星系的起源和演化提供了框架。毕竟，星系对像我们这样的生命形式来说，仍然是非常重要的。

65

物质世界

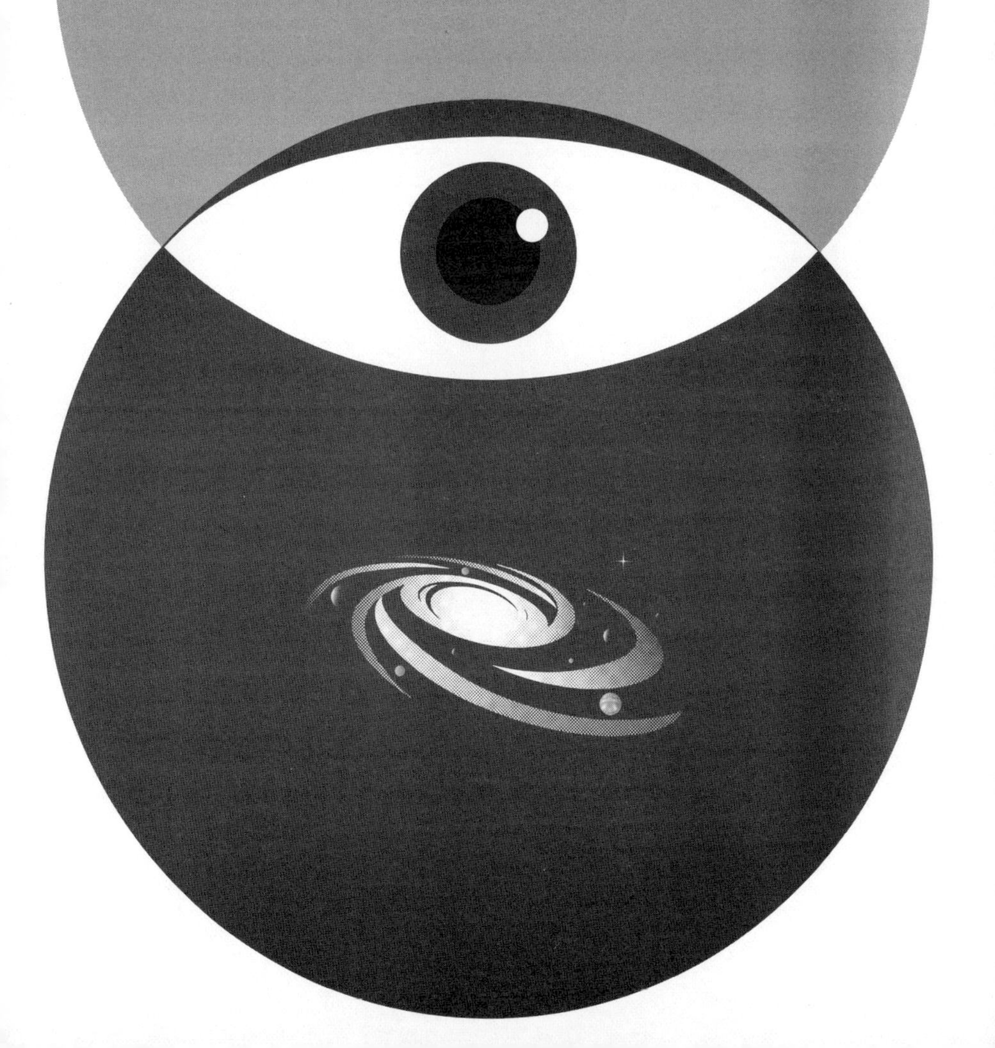

星系是由什么构成的？很明显，星系是由热的、明亮的恒星和冷的、成分是气体和尘埃的暗云构成的。从本质上说，地球和我们的身体都由同样的材料构成，那就是原子。原子由致密的原子核组成，周围环绕着电子云。原子核又由质子和中子组成，原子核中的每一个质子都对应一个电子。在恒星内部，电子被“剥离”出原子，形成一种被称为等离子体的物质，但它们本质上仍然是同一种物质。质子和中子是一个被统称为重子（baryon）的粒子家族的成员，而“重子物质”一词经常被天文学家用来指组成恒星、气体云、行星和人的物质。电子是另一个家族的成员，被称为轻子。但由于电子的质量小于质子或中子质量的千分之一，就质量而言，重子支配着我们熟悉的物质。

66

现代宇宙学的卓越成就之一，就是能够解答宇宙中的重子数量问题。或者更确切地说，是在整个可见宇宙中，重子物质的平均密度应该是多少。依据广义相对论，宇宙学家在测量这类密度时，用希腊字母奥米伽（Ω）命名。在广义相对论中，这个密度与空间

67 的总体曲率有关。这一点其实并不难理解，只要将空间的三维曲度与二维曲度做一个简单的类比即可。地球表面就是二维封闭曲面的一个例子，其自身处处弯曲。在这样一个封闭的表面上，如果沿着相同的方向走足够长的时间，就会回到起点。马鞍面是二维开放曲面的一个例子，可以向四面八方无限延伸。恰好在这两种曲面之间有一种平坦的表面，就像桌面，本身一点儿也不弯曲。爱因斯坦方程表明，三维空间的形状，依赖于所含物质的多少，既可以是封闭的，像地球表面；也可以是开放的，类似于马鞍面；还可以是平坦的，就像桌面。将平坦宇宙所需重子物质平均密度的 Ω 值确定为 1[1]，则封闭宇宙需要的 Ω 值大于 1，开放宇宙需要的 Ω 值小于 1。为简便计算，我们用密度实测值与平坦宇宙的 Ω 值之比来表示宇宙的平均密度。例如，如果宇宙中重子物质的数量是平坦宇宙

68 所需数量的一半（事实并非如此），那么宇宙学家就说，Ω（重子）= 0.5。

宇宙中所有重子物质都是在大爆炸中产生的，都是由纯能量按照质能方程 $E = mc^2$ 的规律转化而来的。当然，这个公式也可以写成 $m = E/c^2$。其实，

1　宇宙学家用没有量纲的纯数来估量重子物质平均密度（Ω）的大小。

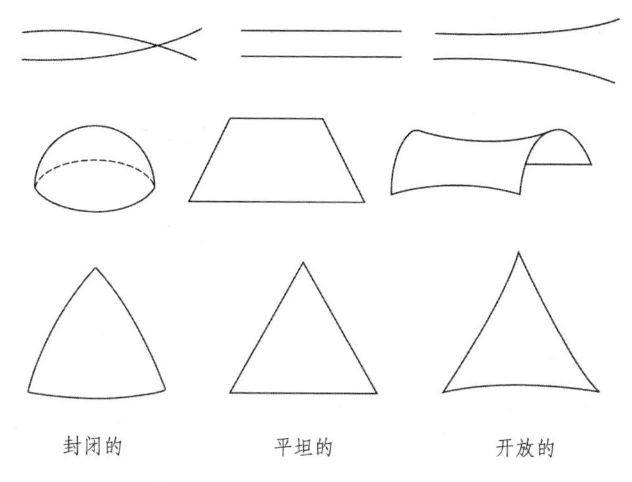

图 14　空间可能符合三种基本几何图形之一。这里是用这些图形在二维中的对等物来表示的

只要有证据能证明大爆炸时的温度至少是 10 亿开尔文，大爆炸中产生的重子物质的数量就可以直接计算出来。证据来自无线电里经常能听到的微弱嘶嘶声。值得注意的是，这种噪声从太空的任何方向都可以接收到。科学家将这些无线电背景噪声解释为大爆炸本身发出的高能辐射传播到现在的剩余辐射。因为可见红移量高达 1000，所以，现在变成了仅比绝对零度高 2.7 开尔文的微波辐射。利用微波背景辐射数据，可以反向计算早期宇宙任何时期的温度。那时的宇宙很小，辐射相应的红移也小。计算显示，大爆炸一秒钟后，温度高达 100 亿开尔文，100 秒后是 10 亿开尔

文，1 小时后冷却到 1.7 亿开尔文，相比之下，太阳中心的温度只有 1500 万开尔文。

在这种条件下，物质以等离子体的形式存在。就像太阳内部一样，辐射在带电粒子之间来回反射。宇宙微波背景辐射本身出现于宇宙开始后的 38 万年。当时宇宙冷却到几千开尔文，大致相当于今天太阳表面的温度。然后，带负电荷的电子和带正电荷的质子形成了中性原子。这使得辐射可以在空间中自由穿行而不受带电粒子的束缚，就像光离开太阳表面向外传播那样。

大爆炸产生了原初火球，其后期的情况与核爆炸内部的情况非常相似。有了对核爆炸原理的了解，宇宙学家可以计算出，大爆炸产生的重子是由约 75% 的氢和 25% 的氦组成的混合物，其中只有极少量的锂。但是，从极端条件下重子粒子与光的相互作用，以及背景辐射的测量数据中计算出的，在大爆炸中产生并存在于宇宙中的重子物质的总量，仅为平坦宇宙所需密度的 4%。换句话说，Ω（重子）= 0.04。

显而易见，研究的下一步是将宇宙中重子物质的预期数量，与在明亮的恒星和星系中看到的重子物质的数量进行比较。基于对恒星的亮度、质量以及恒星在星系中的数量的理解进行粗略计算，可知大约有 1 / 5 的重子物质——也就是不到平坦宇宙所需

69

总量 1% 的物质存在于明亮的天体中。其他 4 / 5 可能存在于恒星之间的气体和尘埃云中，也可能以死亡、燃尽的恒星的形式存在。其中一些是以一种透明的氢气和氦气云雾的形式，存在于像银河系这样的星系周围。然而，正如之前提到过的，从星系的旋转和在空间中移动的方式可知，实际控制星系的物质远比可见物质多得多。这些多出来的物质，只可能是冷的、暗的、非重子的物质，由一种或多种在地球上从未发现过的粒子组成。这些物质被称为冷暗物质，寻找这种物质是目前粒子物理学家最紧迫的任务之一。

冷暗物质存在的证据来自星系的旋转方法及其在太空中的运动方式。圆盘星系的旋转很容易用多普勒效应证实。多普勒效应显示，位于圆盘星系一侧的恒星随着星系的旋转向我们移动，而另一侧的恒星则在远离我们。虽然这种方法只适用于侧面朝向地球的星系，但这类可供研究的星系的数量依然很多。多普勒效应使圆盘星系一侧的红移增加，另一侧的红移减小。因此，利用从星系盘上不同位置测得的红移，就能揭示星系中的恒星围绕星系中心的运动状况。问题在于星系中心核球的外侧，那里有意想不到的事情发生，从核球外侧一直到星系盘的边缘，恒星绕星系中心旋转的线速度居然都是相同的。星系盘中所有的恒

71

星都以相同的线速度运动，这与太阳系的行星绕太阳公转的方式有很大的不同。

图 15　由 WMAP[1] 得到的微波背景图

行星是围绕中心大质量恒星太阳运行的小天体，太阳的引力支配着行星的运动。正因为如此，各行星的运动速度（单位为千米 / 秒）与其到太阳系中心距离的平方根成反比。木星比地球离太阳更远，所以运行速度比地球慢，轨道也更长。但是星系盘中的所有恒星都以相同的速度绕星系中心运动。当然，离中心较远的恒星轨道也较长，所以需要较长的时间来完成围绕星系的一个循环。但星系盘中恒星的轨道速度并无差别。

1　威尔金森宇宙微波背景辐射各向异性探测器（Wilkinson Microwave Anisotropy Probe），于 2001 年 6 月发射升空。

这正是嵌在大量引力物质中的相对较轻物体的轨道运动模式，就像面包里的葡萄干随面包移动一样。自然的结论是，包括银河系在内的圆盘星系，正处于由看不见的暗物质组成的更大的云或晕圈之中旋转（图 16）。这些看不见的东西一定是某种形式的弥散物质，因此必须以粒子形式存在，有点像气体粒子那样。这些物质具有质量和影响普通物质的引力，但不以任何其他方式（例如通过电磁）与普通物质相互作用，否则早就被注意到了。在这个结论所猜测的图景中，冷暗物质粒子无处不在，包括你正在阅读此书的地方。这些粒子不断地穿过你的身体而不会造成任

72

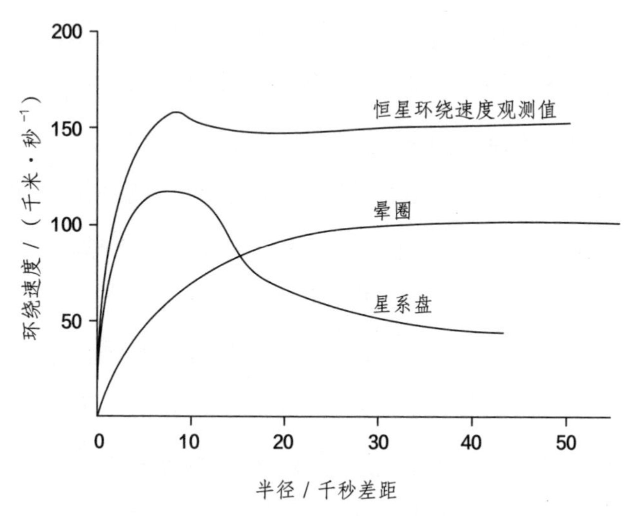

图 16　圆盘星系中典型的“旋转曲线”的示意图

何影响。在每立方厘米里都有数以千计，或许是数以万计的冷暗物质粒子，就连“虚空”，也就是所谓的“真空”里亦是如此。

冷暗物质的存在，也可以通过其对星系团的影响来显示。多普勒频移可以用来揭露星系团中单个星系相对于星系团中心的运动方式，以及星系团中所有星系的速度范围。星系团之所以能够存在，是因为有引力，否则宇宙的膨胀会把它们拉开，并将星系散布到太空中。但这种引力束缚的有效性是有限的，一个抛向空中的球能到达的高度，取决于球的初速度和地球引力的大小，初速度越大，球能达到的高度越高。如果初速度不够大，球就会在地球引力的作用下落回地面。但如果初速度足够大，球就会脱离地球，进入太空。足以使物体脱离地球引力所需要的、垂直于地面的最小速度叫作逃逸速度。逃逸速度只取决于引力天体的质量以及物体离该天体中心的距离。在地球表面，逃逸速度是 11.2 千米 / 秒。对星系团中的一个特定星系来说，通过测定它的亮度，并适当考虑其中暗物质晕的影响，就可以推算出其质量。如果将星系团中所有星系的质量加起来，就可以推算出星系团的逃逸速度。事实证明，为了保持星系团对星系的引力控制，在各星系之间的“真空区”以及单个星系的晕圈

73

中，必须有更多的暗物质。因此，可断定整个宇宙充满了冷暗物质的无形迷雾。

把所有的证据放在一起，就可以计算出宇宙中冷暗物质的数量，它们几乎是重子物质的 6 倍。换句话说，Ω（CDM）= 0.23。再加上宇宙中已知的重子物质，我们就能得出平坦宇宙所需物质的总量是 27%。也就是说，Ω（物质）= 0.27。

这可能会让天文学家感到尴尬，因为当计算精确到这个数值时，也就是大约在 20 世纪末，平坦宇宙的其他证据出现了。证据来自对宇宙微波背景辐射的研究（图 17）。这些宇宙微波背景辐射数据，是由气球和卫星搭载的仪器，从地球大气层的遮蔽之外获取的。相对而言，这些仪器非常灵敏，可以分辨出天空中各处辐射温度的变化，洞察宇宙初期几十万年间留下的冷热印迹。

在宇宙冷却到能形成电中性原子之前，辐射和带电的物质粒子彼此耦合，使得宇宙中不同位置的物质密度差异与辐射温度的差异具有相关性。大爆炸后大约 30 万年，宇宙冷却到临界温度，辐射与物质解耦。在解耦时，辐射保留了与当时重子物质密度变化规律相对应的印迹，有冷有热——那是一种在解耦时重子物质在大尺度分布上的化石。因为光以有限的速度

74

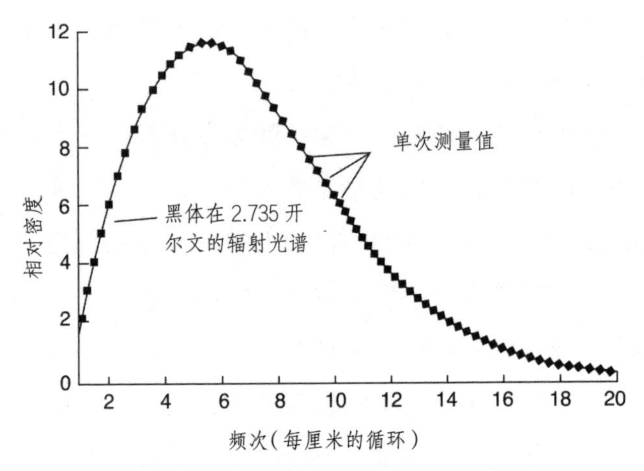

图 17　COBE[1] 卫星测量的宇宙微波背景辐射光谱

传播，所以在 30 万年时，光只能传播 30 万光年的距离。由此得出从大爆炸到解耦的时间里，宇宙中可能存在的任何内部一致性的形式的最大区域，大小都不超过 30 万光年。这意味着，在天空微波背景图中看到的最大的均匀区域，对应着解耦时留下的直径为 30 万光年的斑块。

从那时起，辐射就可以在不与物质直接相互作用的情况下穿过空间。不过，辐射受空间曲率的影响。如前所述，像太阳这样巨大的天体会使经过其边缘的光发生弯曲。这与透镜弯曲光线的方式非常相似。透

1　宇宙微波背景探测器（Cosmic Background Explorer），于 1989 年 11 月发射升空。

镜可以使远处天体的图像看起来更大，比如望远镜；也可以更小，比如把望远镜反过来看。时空弯曲也是如此，这取决于曲率的性质。如果解耦时宇宙的直径是 30 万光年，利用广义相对论，就可以计算出背景辐射中最大的均匀斑块在今天的仪器中看起来应该有多大。观测的结果取决于精确的曲率。如果宇宙是开放的，应该看到一个缩小的斑块；假如宇宙是封闭的，就应该看到放大的斑块；倘若宇宙是平坦的，看到的就是大小不变的斑块。测量结果表明，宇宙几乎可以肯定是平坦的，也有很小的可能是接近封闭的。换句话说，$\Omega = 1$。

然而，宇宙中已知的物质还不到平坦宇宙所需物质总量的 1 / 3。这确实会令人尴尬。但就在宇宙学家刚开始为这个谜题而烦恼的时候，超新星研究出现了，研究表明宇宙的膨胀正在加速。这种加速需要宇宙常数 Λ 有一个确定的值。这个值代表的质量密度大小相当于平坦宇宙所需物质密度的 73%。换句话说，$\Omega(\Lambda) = 0.73$。这正是宇宙学需要的数量。Ω（物质）= 0.27 的发现非但不是尴尬，反而是一场胜利。如果考虑所有因素，就可以得到一个非常简单、非常正确的方程：

75

$$\begin{aligned}\Omega &= \Omega(\text{重子}) + \Omega(\text{CDM}) + \Omega(\Lambda)\\ &= 0.04 + 0.23 + 0.73\\ &= 1\end{aligned}$$

76

正如米考伯（Micawber）先生所说："结果令人欣慰。"由于显而易见的原因，这一系列的成果被称为"ΛCDM"宇宙学，是科学史上最伟大的成就之一。

关于对宇宙的理解的下一个阶段，一项仍在进行中的工作是在"ΛCDM"宇宙学的框架内，解释宇宙中星系的起源。但在能做到这一点之前，需要盘点一下物质世界，即必须解释一下不同种类的星系。不幸的是，圆盘星系和椭圆星系之间并不存在清晰的界限。

像银河系这样的旋涡星系，可见部分由典型的两块构成：一块是星系盘，另一块是中央星系核，尽管有时候星系核非常小。旋臂是圆盘星系最明显的可见特征，但大量的尘埃和气体也同样重要，因为它们为圆盘星系中炽热的年轻恒星的形成提供了原料。这些恒星被称为星族Ⅰ星。星系核和围绕圆盘星系的球状星团中的恒星是较老的星族Ⅱ星。星系中心有没有棒状结构，这可能是旋涡星系某个演化阶段的暂时特征。旋涡星系是最明亮的星系。现在人们普遍认为，在所有圆盘星系的中心都有黑洞，就像银河系中心有黑洞一样。最大的旋涡星系可能含有多达 5000 亿颗恒星。

没有旋臂的圆盘星系，由于历史原因，有时被称为**透镜星系**（lenticular galaxies），它仍然具有基本的星系盘和核球结构，但缺少尘埃云。这些星系大多是由星族Ⅱ的恒星组成的。由此推断，这种星系已经耗尽了所有生产恒星的原料，并进入了沉静的中年时期。从不同角度观看到的遥远透镜星系很难与椭圆星系区分开来，尽管表明其星系身份的特征——旋转，可以通过多普勒效应测出来。

77

椭圆星系总体上并不旋转，但单个恒星会绕星系中心沿椭圆轨道运动。在可以详细研究的邻近椭圆星系中，我们能够分辨出许多在不同方向上沿不同轨道运动的恒星流。这些星流就像银河系中的恒星流一样，但规模更大。这种朝不同方向运动的星流塑造了椭圆星系的整体形状。严格来说，椭圆星系是椭球形的，就像一个被压扁或拉伸的球体。椭圆星系主要由大量古老的星族Ⅱ星构成。从表面看去，它们更像是一个没有星系盘的圆盘星系的星系核。虽然大部分椭圆星系中缺少气体云，但至少有一部分椭圆星系中确实含有尘埃。这些尘埃通常分布在环绕星系中心的环中，但目前这些环中很少有恒星形成。虽然大多数明亮的星系是旋涡形的，最大的星系却是巨大的椭圆星系，包含了超过1万亿颗恒星，直径高达数十万秒差距。但是宇宙中最小的星系似乎也是椭球形的，只有几百万颗恒星，直径通常只有1000秒差距。这些矮星系与最大的球状星团大小相当，这很可能是球状星团起源的线索。我们通常只能在周边看到这么小的星系，离我们最近的几十个星系中有一半是矮椭圆星系。很有可能宇宙中的大多数星系都是这样的矮星系，但是离得太远难以被观测。

任何不能被描述为椭圆星系或圆盘星系的，都被归为不规则星系。不规则星系通常含有大量的气体和尘埃，其中恒星的形成活动非常活跃。因为没有像旋涡星系那样清晰的结构，所以形成的都是松散的恒星集群，点缀在星系周围，使其在照片上呈现不规则的斑驳外观。大小麦哲伦星云是两个受银河系引力控制的小星系，过去被归为不规则星系，但现在人们发现它们有隐藏的棒旋结构，只是由于其恒星具有不均匀性，很难被看清。一些不规则星系很可能是原星系在与更大星系近距离接触时，被更大星系的潮汐力撕裂后留下的残留物或碎片。宇宙中这样的例子比比皆是。这可以从望远镜拍摄的照片中得到证实。一些照

78

片中，可以看到近在咫尺的星系被潮汐力拉伸和扭曲；另一些照片则展示了相互碰撞的星系，也许这两个星系正处于合并的过程中。这些都是不同种类星系起源的重要线索。

星系之间的碰撞也能引发形成恒星的大规模爆发，天文学家将其通俗地称为**星暴**。星暴星系（starburst galaxy）没有正式的定义，但其形成恒星的速率非常高，所有可用的气体和尘埃会在远少于宇宙年龄的时间内耗尽。所以，这种现象不可能持续很长时间。在一些星暴星系中，恒星以每年数百个太阳质量的速度形成，大约是银河系恒星形成速度的 100 倍。这些星系通常会在大约 1 亿年内耗尽所有可用的物质，历时不到宇宙年龄的 1%。

一些星暴星系，尤其是较小的那些，看起来非常蓝，因为它们的光主要是由炽热、年轻的蓝色恒星发出的。这些星系几乎没有尘埃，可能是因为近期曾与另一个星系发生了碰撞或相互作用而受干扰。这种干扰搅动了气体尘埃云，引发了恒星形成的爆发，从而耗尽了储存的气体尘埃。从局部看，单个的星暴发生在星暴星系的星团中。这些致密的星团直径通常在 20 光年左右（6～7 秒差距），亮度是太阳的 1 亿倍。从更大范围看，一些星暴星系非常大，颜色非常红，可以用卫星携带的仪器在红外波段探测到。它们之所以呈现红色，是因为这些星暴星系被大量的尘埃所包围，尘埃吸收了星系中年轻恒星发出的光，并将其重新以红外线形式辐射出来。X 射线望远镜可以透过尘埃，因此我们能知道在这些大的星暴星系中，许多都有两个活动星系核。这表明它们是两个大星系合并的结果。双核星系包含两个黑洞，它们显然分别来自合并前的两个星系。一旦天文学家掌握了寻找星暴的技术，并抱有明确的目的，发现星暴星系就非常容易了。

79

星系中心黑洞的存在也解释了为什么有些星系的星系核会出现剧烈爆发，并把物质抛向太空。这些天体是天文学家在长达几十年的时间里，通过使用不同类型的望远镜，利用电磁光谱的不同频段逐渐发现的。使用的频段包括可见光、射电、红外线、X 射线等。因此，这些天体有过许多不同的称呼。现在认为它们同属于一个家族，其通用名称为“**活动星系核**”（active galactic nucleus，AGN）。这类天体有许多名字，比如塞弗特星系（Seyfert galaxy）、N 星系、蝎虎座 BL 型天体（BL Lac Object）、射电星系和类星体。现在人们普遍认为，这些爆发都是由同样的过程驱动的，包括物质落入（或正在落入）一个超大质量的黑洞，它们的区别不是种类不同，只是程度不同。

当物质落入黑洞时，与之相关的引力能（势能）被释放出来，随着物质加速转化为运动能（动能）。这就像把东西从楼上的窗户扔出去一样，只是事件发生的尺度不同而已。随着势能转化为动能，物体的下落速度越来越快；当下落的物体撞击地面时，动能转化的热能被地面上的分子吸收。当那一小块地面稍微变热时，其中分子的移动速度会加快一点儿。在诸如板球比赛等体育赛事的电视转播中使用的“热点”技术，就是利用这种性质来精确显示球被击中的位置。

80

落向黑洞的物质粒子也会相互碰撞，经过漏斗细细的出口时变得很热，形成由热物质组成的旋涡盘，称为**吸积盘**（accretion disc）。黑洞的引力场非常强大，大量的能量以这种方式释放，最多可达正在下落物质的质量能量（即 mc^2）的 10%。如果星系中心黑洞的质量只有太阳质量的 1 亿倍，或者说大约是星系中所有明亮恒星质量总和的 0.1%，那么它每年只需要吞下几颗相当于太阳质量的恒星，就足以提供在最活跃的星系核中所能看到的能量输出。

所有的大型星系都有可能经历这样一个活动阶段。当所有靠近中心黑洞的“燃料”被吞噬之后，星系就会像银河系一样，安静而体面地稳定下来。但是，如果发生了星系碰撞，引发的扰动使足够多的物质落入吸积盘，比如气体、尘埃，甚至是恒星，黑洞就会被重新激活。任何命中注定要落入黑洞的恒星，在被吞噬之前，都会被潮汐力撕裂成基本粒子。

来自星系中央喷射源的能量喷流，通常沿垂直于星系盘盘面的两个方向射出。这可能是因为黑洞周围物质的吸积盘阻止了能量沿着“赤道”逃逸。因此，物质和能量都可以从星系的中心区域喷射出来，有时会形成稀薄的喷流，与周围环境相互作用，在星系的两侧产生射电噪声瓣。最活跃的星系核，也就是类星体，非常明亮，以至于很难（有时是不可能）在它们的强光下看到周围星系的恒星，因此，类星体看起来就像普通照片中的恒星，只有通过红移测量才能辨识其身份。类星体辐射的能量通常是银河系所有恒星辐射能量总和的 1 万倍。有些类星体的辐射能量甚至可以用地面的光学望远镜观测到，就算它远在 130 多亿光年外，红移大于 6。许多已知的类星体红移大于 4，它们与我们的距离相当于 100 亿光年。但是，类星体过于明亮，不可能是与其周围天体类似的存在。令人高兴的是，最近，科学家将哈勃太空望远镜的能力发挥到了极致，探测到了大量更暗淡、更遥远的天体，更接近大爆炸时相对安静的星系。

81

研究宇宙中远距离天体的重要性在于：当你看到了一个远在100亿光年之外的天体，实际上是看到了这些天体在100亿年之前发出的光。这就是在“回溯时间”，意味着在某种意义上，望远镜就是时间机器，可以向我们展示宇宙年轻时的样子。从这个层面上说，来自遥远星系的光是古老的，经过了漫长的旅程；但是，我们借这些光看到的是年轻时候的星系。早期的研究表明，类星体在宇宙较年轻时更为常见。正如你所预料的那样，类星体靠吸积获得动力，并在吞噬了所有可用物质后逐渐暗淡。从历史上看，这是使主流观点偏向大爆炸模型而非稳态模型理论的证据之一。哈勃太空望远镜所做的最深入的观测是回看130多亿年的历史，我们从中收益良多。

关于这一切，还有一件特别的事情。对遥远的天体来说，由于光经过很长时间才到达地球，所以在光走向地球的过程中，宇宙也显著地膨胀了。本以为回望的时间是4.25年，其实我们看到的应该是4.25光年外的天体；回溯42.5亿年，看到的就应该是42.5亿光年处的物体。但要注意，这其实是光从那些天体出发时间的情况。真实的情况是，那些天体现在的距离要远得多！在42.5亿光年这一情况下，天体远了接近一倍（实际情况甚至比这更复杂，因为光需要走过的距

82

离，从出发时就开始增加了。但这极度简化的结果也足以让人们更加重视这个问题）。这就产生了一个问题，即到一个遥远星系的“当前距离”到底是什么意思，尤其是因为没有什么能比光快，所以我们没有办法测量“当前距离”。因而和其他天文学家一样，我将用回溯时间作为衡量天体距离的关键标尺，而不是试图把它转换成我们所在宇宙区域与任何天体的“当前距离”。本章前面提到的“距离”实际上应该被视为与回溯时间对等的量。

在摄影和电子记录相对于人眼的诸多优势中，最重要的是看的时间越长，看到的东西就越多。人类眼睛的功能是对周围环境进行实时观察，眼睛有很大的局限性，例如只能看到比某一限度更亮的东西，像是恒星。如果一个天体因太暗而不可见，你再怎么盯着那个方向看也无济于事，你的眼睛已经适应了黑暗也不行。但是，安装在现代望远镜上的探测器，只要指向光源，就会不断把微弱的光线累加起来。很明显，长时间的曝光会比短时间曝光更有可能显示出暗淡的物体。因为曝光时间越长，探测器收集到的光子就越多，成像就越清晰。迄今为止，曝光时间的最长纪录是哈勃太空望远镜的100万秒。这个纪录产生于2003年9月24日至2004年1月16日期间，那时天文学家将哈勃太空望远镜对准了天炉座的一小块天空。在普通照片中，这片天空看起来完全是黑色的。这张以数码方式拍摄的照片，是在 110 多天里，通过 800 次曝光得到的。每次曝光后得到的图像以电子方式存储，然后经过计算机合成。合成后的照片相当于一次性曝光超过 11 天的照片（图 18）。这张照片显示，那片看似空荡的天空中充满了星系，并且还捕捉到了一些在宇宙不到 8 亿岁时发出的、红移约为 7 的光线。

83

图 18　哈勃超深场

这张照片被称为哈勃超深场（Hubble Ultra Deep Field，HUDF）。图像覆盖的那片天空只相当于整个天空面积的一千三百万分之一，看上去就像一臂之外的一颗沙粒，天文学家说，这相当于透过一根 2.5 米长的吸管进行的观察。然而在哈勃超深场图像中可以看到，这一小块天空包含了大约 10 000 个星系。特别有趣的是，这些星系中最暗淡、最红的天体，回溯时间最久。来自这些特殊天体的光，以每分钟一个光子的速度进入哈勃太空望远镜的探测器。

84

尽管哈勃超深场中有许多正常的星系，包括旋涡星系和椭圆星系，但距离较远的天体却有着各种各样的奇异形状。其中一些特殊形状显然与彼此之间的相互作用有关。有些星系看起来像手链上的环，有些则像牙签一样又细又长。还有一些由于形状过于奇特，难以描述。这并不奇怪，因为在宇宙历史的早期，没有旋涡星系和椭圆星系，也没有任何类似于我们周围星系的天体。天文学家将此解释为证据，证明他们已经捕捉到了星系形成早期阶段的快照。那时的星系还没有形成现代宇宙中可见的规则结构。当下一代望远镜可以用来回溯更遥远的时间，也就是进入所谓的“黑暗时代”时，不要指望照片上能看到任何东西。这里所谓的黑暗时代，是指从宇宙大爆炸几十万年之后辐射与物质解耦，到大爆炸后几亿年后第一批星系形成之间的那段时间。在这种情况下，什么也探测不到将是对科学理论的成功确认。在哈勃超深场中发现的最古老的天体本身可能正好处于黑暗时代的边缘，大约在大爆炸后的 4 亿年，红移约为 12。

这些星系也许应该称为原星系。最值得注意的是，它们居然在这个时候就已经存在了。在不到 10 亿年的时间里，宇宙从一个充满热气的海洋，变成了充斥着物质团块的地方。那些物质团块大到足以形成后来的星系，并在引力的作用下凝聚。若非如此，物质会伴随宇宙的膨胀弥散开来，变得越来越稀薄。必须有某种能让星系生长的种子存在，其核心得有强大的引力，大到足以克服宇宙膨胀使物质变稀薄的趋势，唯有在这种情况之下，星系生长才可能发生。星系中心超大质量黑洞核心的确认，证明了星系形成模型的最后一个环节，解释了像银河系这样的星系是如何演化而来的，也最终解释了为什么我们在这里——我们是银河系的一部分。

85

07

星系的起源

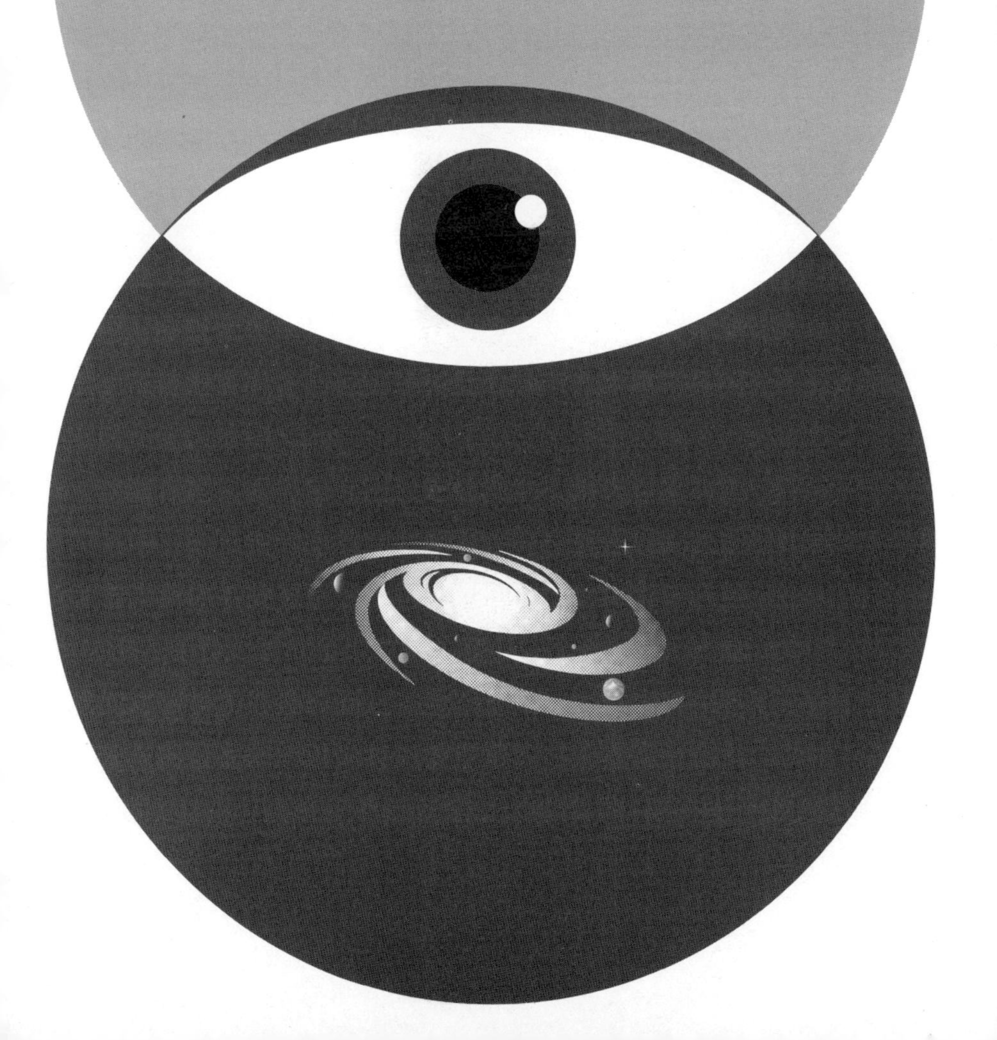

在详细了解星系的起源之前，我们有必要先盘点一下宇宙的主要结构特征，这有助于从总体上准确把握星系起源的脉络。前面的章节描述了单个星系的性质和外观，并提到了一个事实，即大多数星系都被引力凝聚在星系团之中。但其实在星系团之上，宇宙中还存在着另外一个更大的结构，其中隐含着星系起源的重要线索。在这种超大尺度上，星系（严格地说是星系群和小星系团）在宇宙中排成纵横交错的丝状结构，大星系团则位于丝状结构的交叉点处。在丝状结构之间是较暗的区域，星系比较稀少。我们可以用从空中俯瞰夜间地面的场景来打比方——在一个很大的发达地区，如欧洲或北美洲，我们可以看到，被街灯和车灯照亮的道路纵横交错，这就好比宇宙的丝状结构：灯火通明的城市是道路的交会处，这与宇宙丝状结构的节点相似；道路之间的乡村则是一片黑暗，对应丝状结构之间的黑暗区域。两者之间最大的不同之处在于：星系在宇宙中的分布是三维的，就像那种泡沫状结构。最近对地球周边宇宙的红移测量展示了这种结构，测量范围远达红移量 0.5 左右的宇宙深处。 86

与星系团和超星系团不同的是，这些丝状天体结构并非引力作用的结果。用道路来类比，这些丝只是星系像某种物质块相互牵引时移动的路径。但丝状结构的存在，确实揭示了有物质在起牵引作用。

87

到目前为止，由天文学家组成的团队，对星系三维分布的总体格局进行了仔细研究。他们用红移来确定星系的距离，绘制了数百万星系在太空中的分布图。观测这些相对较近的宇宙后，可将结果与微波背景辐射中看到的热点和冷点的模式进行比较（微波背景辐射留下的红移为1000），还可以与在各种不同宇宙模型中模拟宇宙的演化结果进行比较。从理论上讲，在宇宙膨胀早期的原初火球阶段，当重子物质和辐射紧密耦合在一起时，空间里充斥着纵横交错的各种波长的声波，声波的速度快到相对论效应完全不能忽略的量级[1]。在重子物质与辐射解耦之后，正如之前看到的，辐射仍然携带着声波形成的模式印迹，而重子则在引力作用下凝聚成物质团块。通过应用统计技术分析周围宇宙中的星系分布模式，天文学家已经发现了这些声波在物质分布中留下的印迹（即所谓的“声波峰值”）。

1 理论分析表明，在宇宙膨胀早期的火球阶段，声波的速度可能高达光速的60%。

2005 年，使用不同分析方法的两个团队都报告说，在大型三维测量中看到的星系分布统计变化显示了来自大爆炸的声波印迹。从观测上看，一切都是一致的。但是计算机模拟显示，如果假定重子凝聚成团仅靠重子自身的引力，那么自大爆炸至今的这段时间内，从大爆炸原初火球的涟漪中，演化出像我们现在的宇宙这样大的结构是完全不可能的。重点是，从波长角度来看，声波波长很长，但对物质分布的影响很微弱，只是宇宙这片海洋上微弱的涟漪。

88

我们需要一些额外的引力作用并不奇怪，因为前面章节中曾展示过暗物质存在的证据：从单个星系的旋转，到星系团被引力束缚在一起。除此之外，计算机模拟的结果又为暗物质的存在提供了一个完全独立的证据。计算机模拟非常深奥微妙，可以精确地得出宇宙中应该有多少冷暗物质存在。

模拟是在膨胀宇宙的数学模型下进行的。一个典型模型中包含了数目庞大的“粒子”，每个粒子的质量相当于太阳的 10 亿倍。模拟过程中，按照已知的物理定律，计算机跟踪每个粒子在引力影响下的运动，再通过计算推演出宇宙演化过程中的物质分布情况。迄今为止，最大规模的模拟使用了高达万亿个粒子。模拟是从物质与辐射解耦时开始的。首先设定粒

89 子的统计分布为物质在解耦时的分布，然后在宇宙膨胀模式下，按照一系列步骤逐步进行模拟。模拟过程中，可以通过选择不同种类的宇宙学常数、不同数量的暗物质，以及不同的时空曲率值，从而推演出不同的宇宙演化过程。模拟过程的计算量十分巨大，需要耗费大量的计算时间。例如，为了获得图 19 所示的模拟效果图，科研人员动用了包含 812 个处理器、2 TB 内存的 Unix 计算机集群，以 4.2 万亿次 / 秒的计算速度，持续运行了几个星期。最终，模拟产生了 64 张宇宙模型的快照，分别对应自大爆炸至今不同时期的物质分布。

图 19　文中描述的膨胀宇宙中物质分布的模拟结构图。这与观测到的星系分布非常吻合

结果很清楚：从统计学角度看，图 19 的模拟效果图显示的物质分布就如同真实宇宙一样，这也是我选择这张图片的原因。事实上，在各种假设的宇宙模型中，只有一个模拟结果与真实宇宙吻合，这张图片就出自使用这个模型的模拟。依据模拟，如果宇宙是平坦的，从微波背景辐射中观察到的不规则现象开始，到今天宇宙中呈现的星系分布，只能在 130 亿年内产生。模拟中，冷暗物质比重子物质多 6 倍，宇宙常数项贡献了宇宙总质量密度的 73%。当然，这是非常成功的ΛCDM 模型。要形成我们观测到的物质分布的结构，其关键在于：在早期宇宙暗物质密度稍高的区域中，引力会稍大一些，形成引力聚集区。重子物质一与辐射解耦，暗物质就会将附近的重子气体拉入引力聚集区，使引力聚集区处气体云的密度逐渐增高。直到坍缩形成恒星和星系，以泡沫模式遍布整个宇宙。在明亮的丝状体之间的黑暗区域中，重子和冷暗物质的密度仍然几乎相同，只需要在各处产生出小小的（浅的）涟漪，就可以创造使气体云坍缩的条件。换个说法，明亮的细丝可以看成河流，重子则沿着河流流动。这就是单个星系形成的理论框架，当然，天文学家们是如此坚信的。 90

在解耦后不久，即使有暗物质存在，重子物质仍然太热，不会形成太大程度的坍缩。关键在于暗物质是冷的，在密度略高于平均水平的地方，立即开始坍缩。直到大爆炸后约2000万年，也就是相当于红移量大约为100时，宇宙还很平滑，但是冷暗物质粒子已经开始把自己拉进引力束缚的团块，因为团块能够在宇宙向外膨胀时锁住物质。从背景辐射中出现的涟漪开始，经过25～50的红移，暗物质就会形成块状物，其质量与地球近似，但直径大致与太阳系相当。这种球状云的大部分质量都集中在中心附近。以这种方式形成的云之间有强大的引力，足以抵抗住宇宙的膨胀，并形成星系团，星系团再结成更大的星系团。就这样，按照“由小到大”的进程，依次推进。这使得重子物质流向质量集中的地方，形成恒星，然后在细丝的节点形成星系，就这样逐步产生了宇宙中具有丝状结构的“宇宙公路”。

宇宙中出现的第一批明亮天体应该是大质量恒星，质量是太阳的几十到几百倍。这些恒星与今天的恒星非常不同，因为这些恒星的内部只有大爆炸产生的氢和氦，没有任何较重的元素。第一批形成的恒星系统，应该是局部丝状结构的一部分。这种结构逐渐成为更大丝状结构的组成部分，逐层延伸到整个宇宙。而且，当星系团和超星系团在细丝中一起“流

91

动”时，这些系统仍然在发展。这些系统表明，恒星形成区出现在大爆炸后约 2 亿年，每个区域包含的质量是太阳质量的 10 万～100 万倍，直径为 30～100 光年，大小类似于今天银河系中形成恒星的气体尘埃云。但这些“云”主要是由暗物质组成的。

对重子在这样的云中聚集形成恒星的模拟表明，每片云中都形成了类似大型丝状结构的小型丝状结构，物质集中在丝状结构的节点上。随着密度的增加，原子间的碰撞变得越来越频繁，一些氢原子聚集在一起形成氢分子。这一点很关键——氢分子能够发射红外线辐射冷却云层内的气体，氦分子也有冷却作用，但效率较低。正是这种冷却促进了重子气体的进一步坍缩，进而形成原恒星[1]，在某种程度上把重子从暗物质中分离出来。

在今天的恒星形成区，由于存在较重的元素，冷却过程更加迅速。这就是气体云能够坍缩得像恒星形成前一样快的原因。但是，在原始恒星形成云中，一切都发生在更高的温度下，这使第一批恒星形成区还存留有质量高达太阳质量几百到上千倍的云。就像今天的恒星形成一样，这些云很难分裂，每一片云只能

1　恒星演化中的一个过程。

形成几颗恒星（可能不超过3颗）。同时，随着原恒星开始变热，周围的一些物质就被吹走了。

这导致了宇宙中第一批恒星的诞生（但令人困惑的是，科学家沿用银河系中恒星的传统，将其命名为“星族Ⅲ”），它们的质量普遍是太阳的几百倍，表面温度约为10万开尔文，在光谱的紫外线部分辐射强烈。这种充满早期宇宙的辐射至今仍然可见，但是由于红移，已衰减为红外线。用斯皮策太空望远镜可以探测到它们。

92

虽然第一批恒星很明亮，但寿命很短。恒星的寿命与其质量成反比，因为质量大的恒星必须更猛烈地燃烧才能支撑住自身的重量。在短短几百万年内，也就是大爆炸之后只有2亿～2.5亿年的时间里，那些质量是太阳质量100～250倍的恒星，就会结束其生命并发生爆炸，将物质抛散到周围的气体云中。抛散出来的物质中包含宇宙首批重元素，它们也随着爆炸的冲击波散布到周围的气体云中。因为这些元素能大大提高冷却效率，结合冲击波对气体云凝聚的促进作用，使得下一代恒星形成时，恒星形成区域中的气体凝聚更容易，不需要聚集太大的质量便能成型。这导致第一批与现代恒星大小相当的恒星出现。事实上，第二代恒星中的一部分，可能仍然存在于银河系中。据计算，最古老的星族Ⅱ星的年龄超过 132 亿年，是在大爆炸后约 5 亿年内形成的。

质量超过太阳 250 倍的恒星在爆炸死亡后并没有被完全摧毁。相反，这些恒星在其中包含的大部分物质坍缩后形成了黑洞。这些恒星形成于宇宙中物质

图 20　一个处于活跃期的黑洞。一个黑洞为从 M87 星系中心发射出来的喷流提供了动力。虽然在可见光照片中可见（上图），但在红外照片中（下图），喷流要清晰得多

密度最大的时期，因此黑洞之间的距离很可能近到足以发生合并，从而成长为质量更大的天体。没有人确切地知道，今天星系中心的超大质量黑洞是从哪里来的。但似乎至少存在这种可能：第一代恒星遗留下来的黑洞合并，开启了疯狂吞噬周围的物质，从而成长为超大质量黑洞的进程。

对红移在 6.5 左右的类星体的观测表明，质量至少是太阳 10 亿倍的黑洞，早在宇宙诞生 10 亿年时就形成了。这是最好的例子，说明类星体足够明亮，以至于它们发出的光在经过漫长的 130 亿年的传播后，仍能被观察到。从另一个角度看，类星体证实了星系开始出现的时间。模拟表明，一定也有不少较小的黑洞演化成了星系生长的核心，每个黑洞都可能处在含有 1 万亿颗太阳质量的物质晕中。重子物质落入黑洞，释放出势能来为类星体和活动星系核提供动力。同时，恒星则形成于那些重子物质凝聚后，较安静的星系外部区域。但是，模拟也表明，大量地球质量大小的原始暗物质云，应该在混乱中幸存了下来，一直到今天，并且仍然存在于星系周围的暗物质晕中。据估计，仅在银河系的晕圈中就可能有 1000 万亿（10^{15}）个这样的天体。

计算表明，按照前述过程，只要中心黑洞的质量不少于100万颗太阳的质量，就可以在几十亿年的时间内形成一个像银河系这么大的星系。令人高兴的是，观测结果显示，银河系中心黑洞的质量大约是太阳质量的300万倍。至此，一切都匹配。尽管天文学家对第一批星系的形成有一个自洽模型，但仍有很多的问题待解释。如星系中心的黑洞质量与周围星系特性之间就存在着某种神秘的关联。

要知道，对这类超大质量黑洞的研究才刚刚开始，而且只能在相对较近的星系中进行研究。研究方法主要是利用多普勒效应测量星系中心附近恒星轨道运行的速度，从而揭示星系中心巨大天体的存在。直到1984年，第一个超大质量黑洞才被发现。从那时起到20世纪末，哪怕是仅仅发现一个黑洞，也是个大事件，因为那时人们对超大质量黑洞的性质还没有足够的认识。但到2000年，已知的超大质量黑洞已经达到了33个，而且每年还会再发现一两个。这就足够用来了解这些天体与其宿主星系之间的关系了。

在21世纪初，天文学家发现了星系中心黑洞的质量与星系盘中央核球部分质量（在椭圆星系中则是整个星系的质量）之间的关系。这与星系盘本身的属性并不相关，星系盘似乎是随着核球的发展而出现的。由于圆盘星系中心的核球与椭圆星系非常相似，所有原始椭圆星系似乎都可能以同样的方式在黑洞周围发育。但并不是所有的原始椭圆星系都发展成了圆盘星系，这可能是因为缺乏可以形成圆盘星系的原材料。因此，当提到椭圆星系以及圆盘星系的核球的一般属性时，天文学家使用了“球状体”（spheroid）这一术语。

超大质量黑洞的质量是通过测量非常接近球状体中心的恒星的速度来确定的。球状体的质量可以根据其亮度来估计，也可以通过计算大系统多普勒效应的平均值，来计算整个球状体中恒星的平均速度，后一种测法叫作速度色散（velocity dispersion），是一种相当独立的测量方法，可以用来揭示球状体的质量，就像根据星系团中星系的运动揭示星系团的质量一样。把所有的数据放在一起，我们就会发现更大质量的黑洞隐居在更大质量的球状体中。这并不惊人，真正令人惊讶的是，这两个质量之间存在非常精确的相关性——中心黑洞的质量始终是球状体质量的 0.2%。

96

这在整个球状体质量中所占的比例非常小，因此也清楚地表明，黑洞本身并不对球状体中恒星的运动速度负责。从引力角度来说，恒星“注意到”的，只是球状体本身的总质量（即恒星以及恒星之间全部剩余气体和尘埃云的质量之和）。从本质上讲，球状体甚至不知道黑洞在哪儿——如果把黑洞拿走，星系看起来还和原来一样。

虽然这种相关性可以非常简洁地用质量来表示，但更重要的方面在于，在更大质量黑洞周围的球状体中，恒星运动得更快。这表明，在星系形成过程中，重子物质云在暗物质晕中坍缩得更严重。换句话说，黑洞在坍缩更严重的系统中会变得更大，这表明黑洞的成长过程是受坍缩滋养的。黑洞的质量是由坍缩过程决定的。超大质量黑洞首先形成，然后在其周围形成星系，这似乎极不可能，两者一定是在协同进化的过程中一起形成的。最初几百个太阳质量的黑洞作为种子，以丝状结构节点中稠密重子云为原料形成。

关于这种协同进化是如何发生的，还没有具体定论。不过在单体上容易看出，从成长的黑洞中喷出的能量，首先会影响恒星在周围物质中形成的方式。然后在某个关键时刻通过将周围的气体和尘埃云赶走，从而停止黑洞的生长和活动，同时也停止了恒星快速形成的早期阶段。这与对星暴星系的观测相吻合。在这些星系中，每年可以看到星系风携带着多达1000颗太阳质量的物质从中心区域吹出。当这样的星系风持续吹拂的时候，将会在稠密的星际云中引发恒星形成，因为当星系风吹向这些云时，会挤压星际云。在

97

可用物质的 0.2% 被黑洞吞噬期间，大约 10% 的重子物质形成了恒星。

中心黑洞质量和速度色散之间的这种关系，适用于是太阳质量的几百万到几十亿倍的黑洞——跨度高达 1000 倍（3 个数量级）。这种关系也可以跨越整个宇宙，从现在到至少 3.3 的红移，那时宇宙只有 20 亿年的历史。当这种关系被首次发现时，人们认为扁平的没有中央核球的圆盘星系也没有中央黑洞。但在 2003 年，天文学家在没有核球的圆盘星系 NGC 4395 中发现了一个黑洞，质量介于太阳质量的 1 万～10 万倍。与太阳相比，这个黑洞是超大质量的，但与前述天体相比，又是轻量级的。这个星系没有核球，却有恒星较为集中的中心，其速度色散分布意味着黑洞的质量大约是太阳质量的 6.6 万倍。换句话说，速度色散和质量关系与更大系统中的关系吻合。可能所有的圆盘星系和椭圆星系都有中心黑洞，不规则星系或许存在例外。

这种关系也适用于我们自己的银河系，以及近邻仙女座星系（M31）。银河系的中心黑洞质量是太阳质量的 300 万倍，而且有一个小的中央核球；仙女座星系的中心黑洞质量是太阳质量的 3000 万倍，并且有相对较大的中央核球。银河系和仙女座星系之间的整体关系也为我们提供了一条线索，让我们了解在早期宇宙中，星系和中心黑洞形成之后发生了什么。

到目前为止，对于较小的椭圆星系和圆盘星系的起源，已经有了较为合理的解释。但与小椭圆星系不同，大型椭圆星系似乎是通过合并较小的星系形成的（正如我前面提到过的）。目前，银河系和仙女座星系正以数百千米每秒的速度相互接近。这两个星系注定不会发生迎头相撞这种事，但在不到100亿年之后，它们将合并在一起，形成一个巨大的椭圆星系。有证据表明，仙女座星系是在吞噬了一个中等大小的同伴后，才壮大成现在这样的，因为仙女座内部似乎存在双核结构。除此之外，可以预期，两个成熟的圆盘星系的合并将会非常壮观。

正如前文所述，与恒星的直径相比，星系中恒星间距很大。即使两个星系真的正面相撞，发生恒星碰撞的可能性也非常小。两个星系会相互对穿而过，并在引力作用下发生扭曲，星系中恒星的轨道也会随之而改变。但巨大的气体云和恒星之间的尘埃确实会发生碰撞，进而也会有引力的挤压和扭曲，导致类似星暴星系中那样的恒星浪潮的形成。当两个星系相互穿过时，气体和尘埃从星系中喷出，形成物质流，其中可能诞生新的球状星团。相互穿过之后，两个星系相互吸引，围绕着彼此开始纠缠旋转，体验着另一种互动。然而，故事远未结束，随着一次次的缠绕旋转，星系的两个核心逐渐靠近，最后不可避免地合并到一起，两星系也就合并成了一个新的星系。新的星系将没有明显的星系盘，但会有许多星流。这些星流带着过去星系的记忆碎片，在新星系里盘旋流动。两个星系中央的黑洞合并时会释放出巨大的能量，触发星暴活动的最后阶段。经过这个阶段后，巨大的新椭圆星系会安定下来，进入宁静的稳定阶段。这种星系合并的倒数第二个阶段，实际上可以在 NGC 6240 星系中看到。那里有两个相距 1000 秒差距左右的黑洞，在星系的中心沿着终将碰撞的轨道互动。

99

过去人们曾认为，银河系和仙女座星系若要完成合并，需要 50 亿～100 亿年的时间。届时，太阳已经走完了作为明亮恒星的生命历程。但在 2007 年，哈佛－史密森天体物理中心的一个研究小组报告称：计算结果表明，银河系可能在大约 20 亿年后就会开始发生扭曲。希望那时太阳系中还有能作为目击者的智能生命存在。不过，任何这样的目击者都必须拥有极大的耐心，因为即使根据修订后的时间表，合并也要再花 30 亿年才能完成。到那时，衰老的太阳将被挪到距合并星系中心约 30 千秒差距的轨道上，那里大约是目前太阳距银河系中心距离的 4 倍。尽管新时间表的可信度尚待证实，但无论何时发生，最终结果都差不多。

近距离接触也会导致星系的收缩。在星系丰富的星系团中，个体成员（就像群蜂中的蜜蜂）在较强引力作用下移动得太快，无法合并。但在互相掠过的情况下，星系会从对方身上剥去尘埃和气体，甚至是恒星。物质将这样从一个星系释放到星际空间，形成**热气雾**。这些热气雾可以用 X 射线波段探测。最大的椭圆星系往往位于这类星系团的中心，就像一只蜘蛛蛰伏在蛛网的中心，吞噬任何靠近的东西，吃得越来越胖。

在低红移的星系中，约有 1% 积极参与了合并的后期阶段。但与宇宙的年龄相比，这些过程所需的时间太短了。统计数据表明，附近可见的所有星系中，大约有一半是在过去七八十亿年里，由大小相近的星系合并而成的。像银河系这样的圆盘星系，似乎是由更小的亚星系合并形成。从球状体开始，随着时间的推移，经历了一点一滴的吸积过程。前面曾多次提到过恒星流，就是银河系捕获的较小天体的残余物。另一个支持这一观点的证据触及了更远的过去，它来自通过光谱组成的可以很准确地推断出年龄的球状

100

星团。

比氢和氦重的元素在第一批恒星中含量极少，而较年轻的恒星则富含前几代恒星制造的重元素，这一点大家都已经很清楚了。每个球状星团都由年龄相同的恒星组成，这证实了每个球状星团都是由同一团气体尘埃云形成的。但星系团之间的年龄各不相同，说明形成的时期不同。最古老的星系有130多亿年的历史，这与我们对第一批星系形成时间的预测非常吻合。球状星团的年龄各不相同支持了这一观点，即银河系原始核球之外的部分，是由成千上万个较小的气体云组成的，每个气体云都含有多达100万颗太阳质量的物质。每当新的气体云与正在成长的星系相撞时，就会产生冲击波，在气体云中激起涟漪，并在其核心引发恒星爆发，形成新的球状星团。来自云团的大部分物质会受到引力的拖曳，因摩擦而减速，最后成为围绕核球成长的星系盘的一部分。一些球状星团侥幸存活到今天，其余球状星团则由于轨道过于深入星系内部，被潮汐力撕裂。然而计算机模拟显示，若要在现在宇宙年龄之内完成整个凝聚过程，必须满足一个条件——暗物质对总引力场的贡献是重子物质贡献的几倍。没有暗物质，圆盘星系就无法生长；没有暗物质，就不会有任何球状体的种子。

101

在这个自洽的框架内，不规则的小星系被简单地看作宇宙早期遗留下来的碎片。虽然很难看到遥远地方的较小星系，但在解释统计数据时可以考虑到这一点：当允许偏差存在时，观测结果告诉我们，宇宙年轻时，小星系比今天能看到的要多很多。许多小星系不是通过合并变大，就是被更大的星系吞并，这正是我们所期望看到的。在另一极端，今天的宇宙中超过一半的重子物质已经转化为巨大的椭圆星系，其中最大星系的质量是太阳质量的数万亿（10^{12}）倍，相当于 10 个银河系这样的星系的质量总和。这些椭圆星系可以追溯到红移为 1.5 的时候。但光谱研究表明，那时这些星系中的许多成员就已经相当古老了。这就要求这些星系必须在红移为 4 或更早的时候就形成。然而，尽管星系合并最频繁的时代可能发生在 100 多亿年前，但最重要的一点是，这些过程至今仍在继续。星系间仍有相互作用和合并，星系团仍在形成超星系团。从这个意义上说，这一星系构成的宇宙还很年轻，远未成熟。但是星系的最终命运将会怎样呢？

08

星系的命运

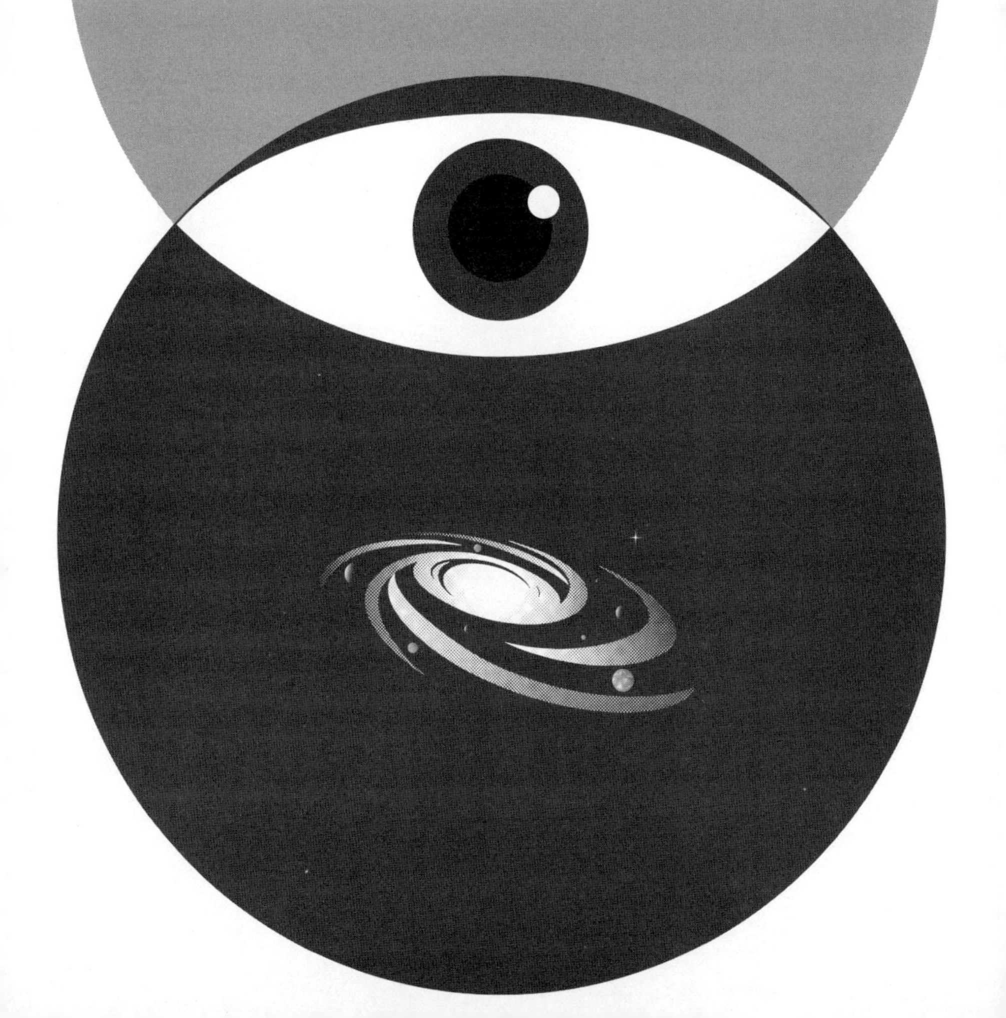

毫无疑问，星系的命运取决于宇宙的未来。不过，就目前来说，宇宙的未来有三种可能的演化方向。尽管学者众说纷纭，但没有人为宇宙命运的三种可能性增添新的内容。宇宙向未来演化的第一种可能性，是宇宙将以与目前大致相同的方式继续匀加速膨胀，也就是膨胀的加速度恒定。虽然已有观测数据的统计分析支持这种可能性，但还不能完全排除其他两种选择。第二种可能性，是宇宙膨胀的加速度会不断变大。第三种可能性，是宇宙将会减速膨胀，或者说，膨胀会在未来的某一时刻停止，并转而开始收缩。在这种情况下，宇宙最终会沿着与大爆炸相反的方向发生**大坍缩**（Big Crunch）。 102

所有这一切都是推测，并且发生在无法想象的漫长时间后，长到时间数字中的零头完全就是累赘。以宇宙当前的年龄，舍去十位以后的数字，保留 100 亿（10^{10}）年作为时间单位才更为简便。除此之外，目前我们对暗物质的性质知之甚少，很难推测暗物质在遥远未来的行为。所以，我们接下来的讨论将集中在重子的命运上，毕竟我们自己就是由这些粒子构成的。

103

如果宇宙膨胀持续足够长的时间，那么最终所有可用的气体和尘埃云都将被耗尽，新恒星的产生也将停止。通过对附近星系中不同星族的观测，以及对本星系中恒星形成速度所揭示的恒星形成史的研究，天文学家推断，当宇宙的年龄达到现在的100倍时，也就是从现在起大约1万亿（10^{12}）年后，就会出现这种情况。随着恒星的死亡和冷却，星系将变得越来越红、越来越暗，星系团也将随着宇宙的膨胀越来越远，届时，任何天文学家都不可能看到宇宙中除本星系群之外的任何东西。在这期间，恒星死亡，然后进入以下三种状态之一。第一种，质量与太阳相当或小于太阳的恒星，其燃料耗尽后的余烬将收缩成白矮星。其中与太阳质量差不多的恒星残余物质，会被压缩进与地球差不多大的球体中。第二种，如果恒星的质量比形成白矮星所需的质量更大，在生命终结时就会进一步收缩，形成更加致密的中子星。中子星的密度非常高，在相当于珠穆朗玛峰体积的球体内，压缩了太阳那样巨大的质量，使得中子星的密度与原子核相当。第三种，如果恒星死亡时遗留下来的质量再大一些，或者，形成中子星后从周围吸积了足够的质量，最终就会坍缩成黑洞。

在这样漫长的时间内，星系也会收缩。原因之一是引力辐射失去了能量。在人类有限的生命内，引力辐射的影响完全可以忽略。但以数万亿年的尺度看，它累积的影响就很可观了。此外，恒星之间的近距离相遇也会导致星系缩小。两颗恒星相遇时，其中一颗获得能量并被抛入星际空间，而另一颗则因失去能量落入离星系中心更近的轨道中。星系团也会以这种方式缩小，最终星系和星系团都会落入由这个过程形成的超大质量黑洞中。

也许，这就是故事的结局了，因为届时宇宙中已经没有可供识别的星系存在了。不过，依然会有黑洞和少量重子物质散落在太空之中。那些重子物质，主要是一些被抛射到星系之外的恒星和少量的气体。但事情并未就此结束。根据粒子物理学理论，如果有足够的时间，就连这些残存的物质也会完全消失。为说明所需时间的跨度，让我们暂时忽略宇宙常数，把目光转向稳定膨胀的宇宙，这个过程更缓慢，缓慢到有无尽的时间思考。 104

从理论角度看，大爆炸中能量转化为物质，但随着宇宙年龄的增长，最终物质也会转化为能量。这里的“最终”是关键词。原子由三种粒子组成：电子、质子和中子。电子是基本的、稳定的粒子；中子在离开原子核后，会在几分钟内衰变成质子和电子，质子在与当前宇宙年龄相当的时间尺度上，似乎是稳定的，但从理论上讲，质子最终也会衰变，每个质子都会变成正电子（相当于电子的反物质）和高能 γ 射线。类似的情况也会发生在白矮星和中子星里的中子上，每次这种衰变都会产生一个电子和一个正电子，以保持电荷的总体平衡。描述大爆炸中物质产生的方程表明，质子的半衰期大约为 10^{32} 年，也就是说，有一半的质子将在 10^{32} 年之内发生衰变。也可以说，在一块含有 10^{32} 个质子的物质中，可能每年都会有 1 个质子衰变。这个数量相当于 500 吨左右的物质所包含的质子数，无论这物质是水、黄油或钢材，任何你能想象的物质都不例外。

这是一段漫长到令人难以置信的时间。即使是 10^{30}，也就是 100 亿的立方，即 100 万亿亿亿年，10^{32} 年也比这还大 100 倍。从现在起到大约 10^{33} 年，如果稳定膨胀能持续这么长时间，届时几乎所有尚未被黑洞吞没的重子，都将衰变成电子、正电子和能量。而且，每当电子和正电子相遇时，就会湮灭成 γ 射线。所以，所有剩下的重子物质最终都会变成辐射。

105

那么黑洞呢？令人难以置信的是，黑洞也不能幸免。广义相对论、热力学和量子理论对黑洞的描述有着深刻的联系。其中的关键是量子物理学的核心原理，即量子的不确定性。不确定性原理指出：在量子世界中，有一些描述粒子属性的物理量，以相关联的方式成对出现，这一对物理量不可能同时拥有精确的测量值。这与测量设备是否存在缺陷无关，而是微观世界的重要特征。能量和时间就是一对这样的物理量。在考虑黑洞命运的时候，重要的是注意能量/时间的不确定性。能量/时间的不确定性表明，没有真正“空”的空间。这说明我们想象中的什么都没有的空间根本不存在，就算再小的空间也不行。这是量子不确定性的要求，在一定的时间内，量子空间可能包含一定量的能量 E。时间 t 越短，能量 E 就越大。因此，从微观角度看，一个小小的能量气泡可以突然凭空出现，然后再骤然消失，而且不会被探测到。由于能量等同于质量，这就意味着，一对粒子，比如电子和正电子，可以从无到有地突然出现，然后迅速消失。

如果这恰好发生在黑洞的边缘，即使两个粒子存在的时间极短，也可能其中一个粒子被黑洞捕获，而另一个粒子逃逸出来。但是宇宙并不会无中生有，因而从整体上看，有一个粒子从黑洞中成功出逃，黑洞的质量因此减小了，而黑洞也缩小了，尽管只是缩小了一点点。利用逃离黑洞表面的粒子，可以给黑洞下一个明确的温度定义，这就是其与热力学有关的原因。相同时间内逃逸粒子越多，黑洞的温度就越高。事实上，小的黑洞温度更高，蒸发得也更快。随着粒子不断逃逸，黑洞内部的剩余质量逐渐下降，当黑洞的质量减少到不足以支撑其与外界的隔绝时，就会以辐射爆发的形式爆炸。黑洞的蒸发过程相当缓慢，质量与太阳相当的黑洞，大约需要 10^{66} 年才会蒸发完毕。而且还需要事先假定，那个黑洞友善到从不吞噬任何外部物质。一个质量相当于一个星系的黑洞将在 10^{99} 年内蒸发，而一个质量相当于一个超级星系团的黑洞——宇宙中有可能形成的最大的黑洞，也将在 10^{117} 年内消失。这实在是极致的推测。

106

话虽这样说，但真有那么长久的时间吗？如果宇宙常数的确是不变的，并且宇宙膨胀的加速度是恒定的，那么除了银河系所在的本星系群之外，一切都将在几千亿年内消失在我们的视野之外。我们所在的本星系团以外的空间，将以比光速更快的速度膨胀。或者说，无论银河系变成什么样子，任何外部信号都永远无法到达银河系。这实际上是说，会有一个缩小的宇宙视界来限定观测的极限。无论视界内外，前述膨胀过程仍将继续下去。事实上，在 10 倍于宇宙当前年龄的时间内，本星系团就将演化合并成一个超星系。但无论这个暗淡的星岛本身如何，都是届时宇宙中可见物质的全部，这个超星系之外的任何天体都不可能被看到。这简直就是天文学预测版的“百思买”，现货现卖，看到的就是可购入的。而且，这里可能有更富戏剧性的结果：假如宇宙常数并不是一个常数呢？

超新星的研究虽然限制了宇宙常数变化的程度，但还不足以证明宇宙常数自大爆炸以来就一直是恒定的。考虑到宇宙常数随时间推移而变化的可能性，也许它真正合适的称呼应该是宇宙学参数。专家从理论角度，就宇宙暗能量密度值的变化将如何影响空间的延展和星系的命运进行了猜测。宇宙常数的变化，可能导致宇宙膨胀的加速度不断增大。这种可能性如果存在，将彻底改变已有的宇宙当前所处发展阶段的结论。这种模型表明，宇宙不是注定长寿的，当前也远非宇宙演化的早期阶段。从大爆炸到一切物质的大终结，可能已经走完了 1 / 3 的路程。更有趣的是，在这种宇宙模型中，如果届时仍有智慧生命的话，他们会活着观察到宇宙[1]最终的毁灭，并且几乎直至毁灭的最

107

1 原文中，用小写的“universe”表示宇宙，因为这些都仅仅是推测，并不代表是现在宇宙的真正成因。这些都是我个人观点，只是猜想，但很有趣！——作者注

后一刻。

宇宙的这种毁灭结局有时被称为**大撕裂**（Big Rip），这样称呼的原因显而易见。撕裂的原因始于这样一种假设，即宇宙膨胀会产生暗能量。然而，如前所述，正是暗能量推动了宇宙的膨胀。如果膨胀本身能产生暗能量，那么宇宙越是膨胀，它所包含的暗能量就会越多，越多的暗能量又会促使宇宙更快地膨胀。如此这般相互促进，直到大撕裂到来。所有这些都不违反已知的物理定律，但也并不是这些定律的必然结果。如果宇宙学参数保持在今天这样较小的水平，那么像太阳、恒星和星系这样的天体，因为引力超过了暗能量的影响，在几千亿年的时间里，都可以毫不费力地抵抗住宇宙膨胀。但在失控的大撕裂场景中，作为一种越来越强大的反引力，暗能量将按指数规律增长，很快就会彻底压垮引力。不仅如此，接下来暗能量会强大到无法对抗，连固体物质都将被无情地撕裂。然而，大撕裂也有“温柔”的一面。按照最极端的大撕裂模型，虽然宇宙将在200多亿年后被彻底撕裂，但一直到最后10亿年之前，像星系这样的物体仍会安然无恙。

108

从现在起到大约200亿年后，暗能量将压垮维系局部星系群的引力，这比宇宙常数恒定时快10倍。

到那时，由银河系和仙女座星系合并而成的大椭圆星系，仍将以可识别的形式存在。尽管太阳已经死去超过 100 亿年，但很可能还有一些智慧生物生活在其他类地行星上，围绕着类似太阳的恒星运行，能够观察到随着宇宙学参数的不断增大所发生的变化。那时的宇宙“视界”仍将保持在大约 70 兆秒差距的大小。

从这一刻开始，用大撕裂前的时间而不是大爆炸后的时间来描述事件的经过，将更方便。在大撕裂前约 6000 万年，暗能量变得强大到足以克服恒星之间的引力，所有星系都会开始解体。但像太阳系这样的行星系统，仍有可能完好无损地在太空中游荡。大撕裂前 3 个月，将行星和其母星维系在一起的引力就会松动。这样继续下去，即便观察者的文明有办法幸免于难，也会在大撕裂前大约半小时，也就是行星被宇宙膨胀撕裂时终结。在最后的几分之一秒，原子和粒子将被撕裂成虚无，留下平坦空荡的时空。大撕裂的某些极端版本表明，一个新的宇宙可能从大撕裂留下的真空中诞生，而我们自己的宇宙可能就是从这样一个真空中诞生的。但如果这种情况是正确的，在大约 200 亿年之后，即在大撕裂之前约 6000 万年，星系的末日就会降临。

109 不过，假设宇宙学参数随着时间的推移而逐渐减小，并且减小到零，我们就会回到永远膨胀的宇宙图景中。那里有不断衰败的物质和持续蒸发的黑洞，这正是开篇中概述的图景。难道这就是结束吗？事实上，宇宙方程并不排除宇宙学参数变为负值的可能性。这也许会从某种程度上使世界末日提前到来，让它几乎和大爆炸离我们的时间一样。但这将是另一种末日——不是大撕裂，而是大坍缩，正如大爆炸的逆过程。

在对大坍缩的讨论中，最好再次使用最极端的方案，因为这个宇宙模型与已知的物理定律相协调，与观测数据相一致。宇宙学参数取正值时，暗能量起着与引力相反的作用，使宇宙膨胀加速；取负值时，暗能量起着引力的作用，将宇宙拉到一起。如果真是如此，宇宙的膨胀可能会逆转。利用迄今为止的观测资料结合理论分析，可以确定宇宙学参数值下降的可能范围。从这个范围可以推算出大坍缩最早可能发生的时间，是在120亿年后，最晚可能远至400亿年以后。和前面一样，这些事件也最好用大坍缩之前的倒计时来描述，同样也可以用可观测宇宙缩小后的大小来表示。因为一切都以同样的方式收缩，即使在视界之外，同样的过程也会同时发生在所有地方。智慧的观察者将不会在现场目睹痛苦死去的宇宙。

要注意的是，宇宙学参数值的变化影响的是空间本身，所以宇宙停止膨胀并开始收缩这个事件，必定是在宇宙各处同时发生的。但是，由于光的传播需要时间，所以无论从宇宙中的任何位置观察，都不会立即看到宇宙的收缩，当然也不会马上得出蓝移的星系主导宇宙的结论。来自附近星系的光会开始蓝移，但是来自遥远星系的光，大部分时间都在膨胀的太空中旅行，仍然会红移。足够长寿的文明能够记录下“蓝移视界”以光速向外扩张的情况，直到蓝移现象最终占据主导地位。

110

就星系而言，在宇宙开始坍缩之后的数十亿年内，几乎感受不到什么影响。前面描述过的恒星形成和星系合并的过程将照常进行。伴随宇宙逐渐收缩，星系群相互靠近并最终合并，并且越来越频繁。但对于生活在像地球这样行星上的生命而言，仍然没有任何重大问题。对生命的威胁，实际上来自目前宇宙最微弱的特征之一：大爆炸留下的背景辐射。

这种宇宙微波背景辐射是宇宙诞生时的原初火球遗留下来的。在大爆炸后30万～40万年发生解耦时，辐射温度就像当今恒星表面的温度一样高，当辐射弥散到日益膨大的空间时，温度一直冷却到

2.7 开尔文（−270.45℃）。但当空间缩小时，辐射就会发生蓝移并压缩，升温的过程与辐射冷却的过程正好相反。当星系团被迫开始合并，将所有星系裹挟其中时，宇宙将只有现在大小的 1%，天空的温度将达到 100 开尔文左右。这还不足以令人担忧。但在接下来的几百万年内，背景辐射的温度将超过冰的熔点（273 开尔文），宇宙中的任何地方都将不再有冰或雪。生命可能仍然存在，但是随着温度的不断升高，温度超过了水的沸点（373 开尔文），随着时间的推移，很快整个天空开始变得越来越亮。

111

在大坍缩之前的几十亿年，生命可能已不复存在，星系被打乱成一堆恒星。在离末日不到 100 万年的时候，重子物质也将不再安全。所有重子物质都在恒星内部“反合成”，分解为原子的两个带电的组件。现在，物质和辐射重新耦合在一起。这与大爆炸后发生的解耦正好相反，二者发生的时间恰好对称，一个是在末日前，一个是在诞生后，都是 30 万～40 万年。不同之处在于，恒星，或者说至少是恒星的核心，仍然可以在这个火球中生存，直到宇宙缩小到其目前大小的百万分之一，且温度超过 1000 万开尔文，与恒星内部的温度相当。然后，甚至是恒星的核心也会在火球中熔化。最终，一切都消失在一个奇点中，就像黑洞中心的奇点，或是宇宙诞生的地方。

这就引出了一个有趣的推测：宇宙的诞生，可能恰恰来自先前宇宙的坍缩，或者说来自我们这个宇宙的前世。这样的宇宙可以沿着相似的循环反复膨胀、坍缩再膨胀。然而，这些都与宇宙中星系的命运无关。在大坍缩的情况下，星系将会在大坍缩前大约10亿年，也许距今只有大约110亿年，被彻底摧毁。

大撕裂和大坍缩都只是推测，我们讨论它们的主要目的是表明可能发生的最极端情况。据我们所知，宇宙不可能在不到120亿年的时间里再次坍缩，在接下来的200亿年里，大撕裂也不会撕裂星系。30年前，天文学家估计大爆炸发生的时间为120亿～200亿年之前，几乎与前面的几种推测同样不确定。然而，目前宇宙的年龄已被更精确地确定在137亿年[1]，这就是进步。也许可以希望在接下来的30年里，人们对宇宙命运的理解也能取得类似的进展。 112

目前对星系命运“百思买”般的预测是：大体上，宇宙常数是恒定的。虽然由于宇宙膨胀在逐渐加速，缓慢撕裂最终可能发生，但这是距今最远的未来，我们最不需要担心。在这种情况下，星系在数千

1　最近更新为138.2亿年。

亿年，也就是目前宇宙年龄的 10 倍以上的时间内，都是安全的。未来的智慧生命将有足够的时间精确地计算出宇宙终结的方式。

全书完

名词对照

X 射线望远镜	x-ray telescope
Λ 场	lambda field

A

矮星系	dwarf galaxy
暗能量	dark energy
暗物质	dark matter

B

白矮星	white dwarf
不规则星系	irregular galaxy
不确定性原理	quantum uncertainty

C

超大质量黑洞	supermassive black hole
超新星	supernovae

D

大辩论	Great Debate
大撕裂	Big Rip
大坍缩	Big Crunch
电子记录法	electronic recording methods
动能	kinetic energy
多普勒效应	Doppler effect

F

分光光度法	spectrophotography

H

哈勃常数	Hubble’s Constant
哈勃超深场	Hubble Ultra Deep Field (HUDF)
哈勃太空望远镜	Hubble Space Telescope
恒星消光	interstellar extinction
红移	redshift
活动星系核	active galatic nucleus (AGN)

J

解耦	decoupling
静态宇宙模型	static universe theory

K

克努特·伦德马克	Lundmark, Knut
空间曲率	curvature of space

L

蓝移	blueshift
类星体	quasar
冷暗物质	Cold Dark Matter (CDM)
猎户座星云	Orion nebula

M

马尔姆奎斯特偏差	Malmquist bias

N

能量 / 时间的不确定性	energy/time uncertainty

P

平庸原则	principle of terrestrial mediocrity

R

热力学	thermodynamics
人马座	Sagittarius constellation
人马座 A	Sagittarius A
人马座矮椭圆星系	Sagittarius dwaf elliptical galaxy
人马座星流	Sagittarius star stream

S

三角视差法	triangulation
射电望远镜	radio telescope
射电噪声	radio noise

声波峰值	acoustic peak
时空曲率	curvature of spacetime
视差效应	parallax effect
速度色散	velocity dispersion

T

逃逸速度	escape velocity
透镜星系	lenticular galaxy
退行速度	recession velocity
陀螺仪	gyroscope
椭圆星系	elliptical galaxy

W

微波背景辐射	microwave background radiation
稳态模型	steady state model

X

仙女座星系	Andromeda galaxy (M31)
仙女座星云	Andromeda Nebula (M31)
小麦哲伦星云	Small Magellanic Cloud (SMC)
新星	nova
星暴星系	starburst galaxy
星云	nebulae
星族	Population
旋涡星系	spiral galaxy

Y

银河系	Milky Way
宇宙常数	cosmological constant
宇宙岛理论	island universes idea
宇宙的坍缩	collapse of universe
宇宙学参数	cosmological parameter
原星系	proto-galaxy
圆盘星系	disc galaxy

Z

造父变星	Cepheid
正电子	positron
质子	proton
中子	neutron
重子物质	baryonic matter

英文原文

John Gribbin

GALAXIES

A Very Short Introduction

For my brother, who suggested I write it.

Contents

List of illustrations

The publisher and the author apologize for any errors or omissions in the above list. If contacted they will be pleased to rectify these at the earliest opportunity.

Introduction

The scientific investigation of galaxies began only a little more than the 'three score years and ten' of a biblical lifetime ago, in the 1920s, when it was first established that many of the fuzzy blobs of light seen through telescopes are islands in space made up of vast numbers of stars, far beyond the boundaries of the Milky Way, our own island galaxy. Without telescopes, we would never have been able to explore the Universe beyond the Milky Way and investigate the nature of galaxies, but it had taken nearly four hundred years for telescopes to be developed to the point where the true nature of galaxies became clear.

As far as anyone knows, the first person to use a telescope to look at the night sky was Leonard Digges, an Oxford-educated mathematician and surveyor, who invented the theodolite some time around 1551. He kept his use of the telescope (essentially, a theodolite pointed upwards) secret, because of the value of the theodolite in his work, but he wrote one of the first popular books of what would now be called science in English. This included a description of the Ptolemaic Earth-centred model of the Universe. Leonard died in 1559, but his son, Thomas Digges, carried on where his father left off. Born in the 1540s, Thomas became a mathematician, and in 1571 arranged posthumous publication of a book by his father in which a telescope was described for the first time in print. Thomas Digges also made astronomical observations, and in 1576 published a revised and expanded version of his father's first book, which included the first printed description in English of the Sun-centred Copernican model of the Universe.

In that book, *Prognostication Everlasting*, the younger Digges said that the Universe is infinite, and included an illustration of the Sun, orbited by the planets, at the centre of an array of stars extending to infinity in all directions. Since we know Digges had at least one telescope, the natural inference to draw is that he had used one to look at the band of light across the sky known as the Milky Way, and had discovered that it is composed of uncountable numbers of individual stars.

The story of Leonard and Thomas Digges may come as a surprise, since the person usually credited both with making the first use of an astronomical telescope and with the discovery that the Milky Way is made of stars is Galileo Galilei, at the end of the first decade of the 17th century. In fact, the telescope was invented independently several times in north-west Europe, and news of the invention only reached Italy, from the Netherlands, in 1609. With only a description of the instrument to go on, Galileo built one of his own – the first of many – and among other things turned it on the heavens. His discoveries were published in a book titled *Sidereus Nuncius* (*The Starry Messenger*) published in 1610. This made him famous, and is the source of the popular myth that he was the first astronomer to use a telescope. But, like Thomas Digges before him, Galileo did indeed observe that the Milky Way is made up of a myriad of stars.

The next step towards an understanding of our place in the Universe was made by Thomas Wright, an English instrument maker and philosopher, in the middle of the 18th century. But, like that of the Digges, his contribution was almost forgotten. The Milky Way forms a band of light across the night sky, and in his book *An Original Theory or New Hypothesis of the Universe*, published in 1750, Wright suggested that it is made up of a slab of stars, which he likened to the shape of the grinding wheel of a mill. Even more impressively, he realized that the Sun is not at the centre of this disc-shaped slab of stars, but out to one side. He even suggested that the fuzzy blobs of light visible though telescopes, known as nebulae from their resemblance to clouds, might lie outside the Milky Way – although he did not make the leap of imagination that would have been required to suggest that the nebulae might be other star systems like the Milky Way itself. Immanuel Kant, another philosopher-scientist, picked up these ideas from Wright; he did take the extra step, and suggested that the nebulae might be 'island universes' like the Milky Way. The idea was not taken seriously.

As telescopes were improved, more and more nebulae were discovered and catalogued. One reason for the careful cataloguing was that astronomers of the late 18th and early 19th centuries were eager to find comets, and at first sight the fuzzy blob of a nebula looks like the fuzzy blob of a comet. So people like Charles Messier, in the 1780s, and William Herschel – who completed a catalogue in 1802 – identified the positions of nebulae in order that there should be no confusion. Herschel's catalogue included 2,500 nebulae, most of which we now know to be galaxies. Over the next 20 years, he tried to find out what the nebulae were made of, but even his largest telescope, with a mirror 48 inches (1.2 metres) in diameter, was unable to resolve the fuzzy patches of light into stars. He died in 1822 convinced that the nebulae really were diffuse clouds of material within the Milky Way.

The next observational step was made by William Parsons, the third earl of Rosse, who built an enormous telescope with a mirror 72 inches (1.8 metres) across in the 1840s. With this instrument, he found that many of the nebulae have a spiral structure, like the pattern of cream stirred into a cup of black coffee. Over the following decades, some nebulae were firmly identified as glowing clouds of gas within the Milky Way, and some were resolved into clusters of stars, on a much smaller scale than the Milky Way and associated with the Milky Way. But the spiral nebulae did not fit either category. The development of astronomical photography in the second half of the 19th century made it easier to study spiral nebulae, but the photographs were not good enough to reveal their true nature.

At the beginning of the 20th century, the vast majority of astronomers agreed that spiral nebulae were swirling clouds of material surrounding a star in the process of formation, like the cloud from which the Solar System was thought to have formed. But over the next two decades the island universes idea gained enough supporters to encourage the US National Academy of Sciences to host a debate on the subject, with Harlow Shapley, then of the Mount Wilson Observatory in California, speaking for the majority view against the island universe idea, and Heber Curtis, of California's Lick Observatory, speaking for it. Held on 26 April 1920, this became known to astronomers as 'The Great Debate'. Although it failed to resolve the issue, it marked the moment when the modern scientific investigation of galaxies began.

Chapter 1
The Great Debate

There were two aspects to the great astronomical debate of 26 April 1920: the size of the Milky Way Galaxy, and the nature of the spiral nebulae. In fact, it wasn't really a debate at all; the two speakers each made presentations 40 minutes long, and there was a general discussion afterwards. The theme of the meeting, held at what was then the US National Museum, and is now the Smithsonian Museum of Natural History, was 'The Scale of the Universe'. Shapley and Curtis had quite different views on what this meant, which they each elaborated on in a pair of scientific papers published the following year. In essence, Shapley thought that the Milky Way *was* the Universe, or at least the most important thing in the Universe, and was interested in the size of our own Galaxy; Curtis thought that the spiral nebulae were galaxies like our own, and was interested in the scale of things beyond the Milky Way.

The debate happened at the time it did because astronomers had recently developed techniques for measuring distances across the Milky Way. Distances to nearby stars can be measured using the same sort of surveying techniques with which Leonard Digges would have been familiar, starting with triangulation. If a nearby star is observed on nights six months apart, when the Earth is on opposite sides of its orbit around the Sun, the star will seem to shift slightly against the background of distant stars. This parallax effect is just like if you hold a finger up in front of your face and close each of your eyes in turn. The finger seems to move relative to the background, and the closer it is to your eyes the bigger this parallax effect is. The size of the stellar displacement and the diameter of the Earth's orbit (itself known from triangulation within the Solar System) are all you need to work out the distance to the star.

Unfortunately, most stars are far too far away for this effect to be measurable. Even the nearest star, Alpha Centauri, is so far away from the Sun that light takes 4.29 years to travel across the intervening space (so it is 4.29 light years away). By 1908, only about a hundred stellar distances had been measured in this way. Other geometrical techniques, based on the way with which stars in nearby clusters are seen to be moving together through space, make it possible to measure distances out to about 100 light years, or, in the units preferred by astronomers, about 30 parsecs (one parsec is almost exactly 3.25 light years). This was just enough for them to be able to calibrate the most important distance indicator in astronomy.

To put the importance of this new distance indicator in perspective we have only to look at the best estimate of the size of the Milky Way made in the early years of the 20th century. The Dutch astronomer Jacobus Kapteyn counted the number of stars visible on same-sized patches of sky in different directions from us, and included estimates of the distances to the stars, based on the techniques I have described and in part on how faint the stars seem from Earth. He inferred that the Milky Way was shaped rather like a discus, about 2,000 parsecs (2 kiloparsecs) thick in the middle, 10 kiloparsecs in diameter, with the Sun somewhere near the centre. This estimate is far too small, we now know, chiefly because there is a great deal of dust between the stars, which Kapteyn did not know, and this acts like a fog to limit how far we can see across the plane of the Milky Way – a phenomenon known as stellar extinction. Just as a traveller lost in fog seems to be alone at the centre of their own small world, so Kapteyn was lost in the fog of the Milky Way and seemed to be at the centre of his own small universe. Less than a hundred years ago, most astronomers thought that this discus of stars essentially represented the entire Universe.

Things began to change in the second decade of the 20th century. Henrietta Swan Leavitt, working at the Harvard College Observatory, discovered that a certain family of stars, known as Cepheids, vary in brightness in a way which makes it possible to use them as distance indicators. Each Cepheid brightens and dims in a regular way, repeating the cycle exactly again and again. Some run through the cycle in less than a day; others take as long as a hundred days. Polaris, the northern pole star, is a Cepheid with a period close to four days, although the brightness changes in this case are too small to be detected with the naked eye. Leavitt's great discovery was that brighter Cepheids take longer to run through their cycles than fainter Cepheids. Even better, there is an exact relationship between the period of a Cepheid and its brightness. A Cepheid that takes five days to run through its cycle, for example, is ten times brighter than one which takes eleven hours to run through its cycle.

Leavitt made this discovery by studying the light from hundreds of stars in a nebula called the Small Magellanic Cloud (SMC), a star system associated with the Milky Way. She didn't know how far away the SMC was, but that didn't matter because all the stars in it are at essentially the same distance from us. So their relative brightnesses could be compared without having to worry if one star looked fainter than another simply because it was farther away. In 1913, the Dane Ejnar Hertzprung measured the distances to 13 nearby Cepheids using geometrical techniques, and used observations of these stars combined with Leavitt's data to work out the true brightness of a hypothetical standard Cepheid with a period of exactly one day. Armed with this calibration, it was possible to measure the distance to any other Cepheid by working out its true brightness from Hertzprung's calibration and its period, then comparing this with how faint the star appeared on the sky – the fainter it was, the farther away it must be, in a precisely calculable fashion. Among other things, this calibration of the Cepheid distance scale meant that the SMC is at least 10 kiloparsecs (kpc) away. Hertzprung's estimate has since been revised in the light of better observations and an understanding of stellar extinction, but in 1913 it marked a dramatic increase in scale from Kapteyn's estimate of the size of the entire Milky Way (the entire Universe!) to suggest that the SMC was so far away.

It was Harlow Shapley who used the Cepheid technique to map the size and shape of the Milky Way Galaxy itself, after having carried out his own calibration of the brightnesses of these variable stars. This was at the heart of his contribution to the Great Debate.

The key to Shapley's survey of the Milky Way was that he was able to use variable stars to measure distances to star systems known as globular clusters. As their name suggests, globular clusters are spherical star systems. They may contain hundreds of thousands of individual stars, and at the heart of such a cluster there may be as many as a thousand stars packed into a single cubic parsec of space – very different from our region of the Galaxy, where there is *no* star as close as 1 parsec to the Sun. Globular clusters are seen above and below the plane of the Milky Way. By measuring the distances to them, Shapley found that they are distributed in a spherical volume of space centred on a point in the direction of the constellation Sagittarius but thousands of parsecs away from us, in the middle of the band of light known as the Milky Way. The implication is that this point marks the centre of the Milky Way Galaxy, and that the Solar System is, therefore, located far out towards the edge of the Galaxy. By 1920, Shapley had come up with an estimate that our Galaxy is about 300,000 light years (nearly 100 kpc) across, with the Sun about 60,000 light

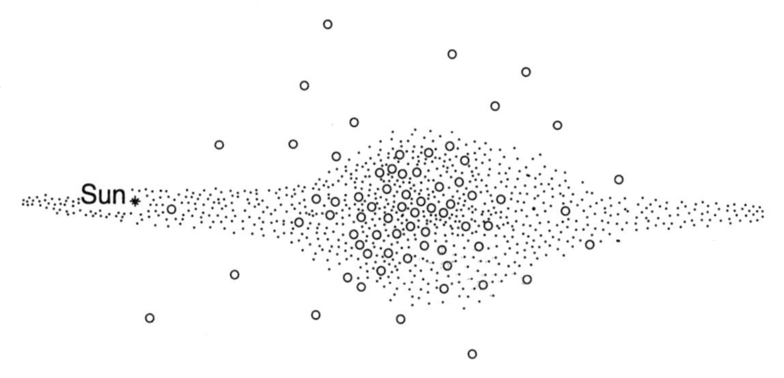

1. The distribution of globular clusters (circles) on one side of the sky implies that the Sun is far from the centre of the Milky Way

years (nearly 20 kpc) out from the centre. As he put it at the Washington meeting:

> One consequence of the cluster theory of the galactic system is that the sun is found to be very distant from the center of the Galaxy. It appears that we are near the center of a large local cluster or cloud of stars, but that cloud is at least 60,000 light years from the galactic center.

On this picture, it seemed to Shapley and like-minded astronomers that the spiral nebulae could not be other galaxies like the Milky Way. Their reasoning was simple. The apparent (angular) size of an object on the sky depends on its actual linear size and its distance from us, in exactly the same way that a real cow standing far away on the other side of a field looks the same size as a child's toy cow held up in your hand. If the spiral nebulae were also about 300,000 light years across, their tiny angular sizes on the sky would place them at distances of many millions of light years, which just seemed too big to take seriously. Instead, Shapley argued that the spiral nebulae were either star-forming systems within the Milky Way Galaxy, or at most small satellites of the Milky Way – islands compared with the continent of the Milky

Way. 'I prefer to believe', he said, 'that they are not composed of stars at all, but are truly nebulous objects.'

He had one other piece of ammunition. Adriaan van Maanen, a Dutch astronomer who happened to be a good friend of Shapley, claimed that he had measured the rotation of several spiral nebulae, by comparing photographs taken several years apart. The effect was incredibly small. In one case, the nebula M101, he said he had measured a displacement of 0.02 seconds of arc, about 0.001 per cent of the angular size of the Moon as seen from Earth. Any such rotation can be converted into a linear speed corresponding to the distance of any part of the nebula from the centre of rotation. This, of course, depends on the actual size of the object that is rotating. If the spirals were the same size as the Milky Way, van Maanen's measurements would have implied speeds comparable to, or even faster than, the speed of light. If he was right, the spirals had to be small objects, relatively close to us. Most astronomers found it hard to believe that van Maanen could actually be making such precise measurements. Later studies showed that van Maanen had made a mistake – nobody quite knows how – but at the time of the Great Debate it was a matter of faith whether you believed his data or not; and Shapley trusted his friend. In his paper published in 1921, Shapley emphasized that van Maanen's results 'appear fatal' to the island universe idea. 'Bright spirals cannot reasonably be the excessively distant objects required by the theory.'

Curtis didn't trust van Maanen's results, and he also didn't trust the still-new Cepheid distance scale. At the Washington meeting, he gave a summary of various earlier estimates of the size of the Galaxy, including, rather cheekily, an estimate for the diameter of just 20,000 light years made by Shapley in 1915. He concluded that 'a maximum galactic diameter of 30,000 light years will be assumed as representing sufficiently well the older view; it is perhaps too large'. This estimate was exactly one tenth of the size Shapley suggested in 1920. Curtis also said that the Sun is located 'fairly close to' the centre of the Galaxy, but not exactly at the centre. But all of this was, to him, a minor matter which he mentioned briefly before going on to discuss the aspect of the story that really interested him, the nature of spiral nebulae and their distances from us.

There were two key facts that Curtis used in his argument that the spiral nebulae are galaxies like our own at great distances from us. The first was the discovery, made by Vesto Slipher, of the Lowell Observatory, that by far the majority of spiral nebulae seemed to be receding from us at very high velocities. The discovery was made by measuring the extent to which lines in the spectra of these nebulae are displaced towards the red end of the spectrum, compared with lines in the light from nearby stars and hot objects on Earth.

Light from any hot object, including the Sun and stars, can be spread out using a prism to make a rainbow pattern, or spectrum. Each chemical element – hydrogen, carbon, and so on – produces a characteristic pattern of bright lines in the spectrum, as distinctive as the bar codes on supermarket goods. When the object is moving away from us, the whole pattern of lines is displaced towards the red end of the spectrum, by an amount which depends on how fast the object is receding; this is the famous redshift. Similarly, when an object is moving towards us, the pattern of lines is shifted towards the blue end of the spectrum – a blueshift. Stars moving around in the Galaxy show both redshifts and blueshifts, corresponding to velocities relative to us of anything from zero to a few tens of kilometres per second.

In the second decade of the twentieth century, measuring the positions of lines in the faint spectra of light from spiral nebulae was pushing photographic techniques to the limit. It was only in 1912 that Slipher was able to obtain such spectrographs of the Andromeda Nebula, also known as M31, which we now know to be the nearest spiral to the Milky Way. He found a shift towards the blue end of the spectrum, indicating that the nebula is rushing towards us at 300 kilometres per second. This was by far the highest such speed measured up to that time. By 1914, Slipher had similar spectrographs for 15 spirals. Only two, including M31, showed a blueshift. The other 13 all showed redshifts, two of them corresponding to velocities of recession of more than a 1,000 kilometres per second. By 1917, he had 21 redshifts, but still only two blueshifts – even today, there are still only two blueshifts. Whatever the nature of spiral nebulae, Slipher's measured velocities implied that they could not be part of the Milky Way; they were simply moving too fast to be gravitationally bound to our Galaxy. Although in 1920 nobody could explain the cause of these large recession velocities, Curtis saw this as evidence that the spiral nebulae had nothing to do with our Milky Way, but were 'island universes' in their own right.

The other main plank in his platform concerned observations of stars that suddenly flared up in bright outbursts. Such stars are known as novae, from the Latin word for 'new', because when they were first observed they literally seemed to be new stars, shining brightly where no star had been noticed before. It is now clear, however, that all novae are outbursts from stars which had previously been leading a quiet life and were too faint to be seen. They are a natural, but fairly rare, stellar phenomenon.

In 1920, Curtis pointed out that 'within the past few years some twenty-five novae have been discovered in spiral nebulae, sixteen of these in the Nebula of Andromeda, as against about thirty in historical times within our own galaxy'. The sheer number of novae seen in the Andromeda Nebula meant that the nebula must be made up of a huge number of stars, assuming that a star in Andromeda was no more likely to become a nova than a star in the Milky Way, and roughly speaking the apparent brightness (faintness) of the novae seen in various spirals was about what you would expect if they were actually as bright as novae in the Milky

2. A classic example of a disc galaxy. This is NGC 4414, viewed by the Wide Field Planetary Camera 2 (WFPC2) on the Hubble Space Telescope

Way, but as remote as the distances implied if the spiral nebulae were the same size as Curtis's estimate of the size of the Milky Way.

There was one fly in the ointment. In 1885, in the very decade that the Andromeda Nebula was identified as a spiral, a bright star flared up in it. The apparent brightness of this nova was about the same as the apparent brightness of a typical nova in the Milky Way. This meant that either the nebula really was part of the Milky Way, or, if the nebula was as far away as Curtis thought, that this was some kind of super-powerful nova, as bright as a billion Suns put together and far brighter than any nova observed in the Milky Way in the 19th century. This was a difficulty for Curtis, which he essentially circumvented by suggesting that there might be two kinds of nova, one much brighter than the other. This seemed like a fudge to his audience at the time; but we now know that there really are stellar outbursts that bright. They are called supernovae, and they can briefly shine as brightly as a hundred billion Suns – as bright, indeed, as all the other stars in a galaxy put together.

As Curtis summed the argument up:

> The new stars observed in the spirals seem a natural consequence of their nature as galaxies. Correlations between the new stars in spirals and those in our Galaxy indicate a distance ranging from perhaps 500,000 light years in the case of the Nebula of Andromeda, to 10,000,000, or more, light years for the more remote spirals... At such distances, these island universes would be of the order of size of our own Galaxy of stars.

In the paper published in 1921, he went further:

> the spirals, as external galaxies, indicate to us a greater universe into which we may penetrate to distances of ten-million to a hundred-million light-years.

In so far as there was a debate on the scale of the Universe in Washington on 26 April 1920, nobody won. Both participants were sure they had come out on top – a sure sign that neither of them had – but both were right on some points and wrong on others. Most importantly, Shapley was right to trust the Cepheid distance scale, even though it hadn't quite been perfected at the time, and Curtis was right that the spiral nebulae are other galaxies. Shapley was also right in placing the Sun far out from the centre of the Milky Way. As for the size of the Milky Way, the best current estimates give a diameter of about 100,000 light years, three times bigger than Curtis's estimate and one third the size of

Shapley's estimate, so by that reckoning they were equally wrong. This does indeed make the Milky Way an average spiral – just how average, I shall discuss in Chapter 4. Although the Great Debate was inconclusive, the key issues it raised were resolved before the end of the 1920s, largely thanks to the work of one man, Edwin Hubble.

Chapter 2
Stepping stones to the Universe

The main reason why the study of galaxies took off in the 1920s was the invention of bigger telescopes and improved photographic techniques, which made it possible to obtain more detailed images (and spectra) of faint and distant objects. Spectrophotography was vital to the discovery of redshifts in the light from spiral nebulae, and photography itself was a key element in the discovery of the Cepheid period–brightness relation. In 1918, a telescope with a 100-inch (2.5-metre) diameter mirror became operational on Mount Wilson in California; it would be the most powerful telescope in the world for almost three decades, and was used by Edwin Hubble to measure the distances to galaxies in a series of steps out across the Universe.

Hubble cut his teeth as a research astronomer as a Ph.D. student at the Yerkes Observatory (part of the University of Chicago) between 1914 and 1917. His research project there was to obtain photographs of faint nebulae using a 40-inch (1-metre) refracting telescope. This was one of the best telescopes in the world at the time, and the largest refractor ever built. By and large, for telescopes the same size, refractors, which use lenses, are more powerful than reflectors, which use mirrors; but reflectors can be made bigger because their mirrors can be supported from behind without blocking out any light. This observing programme led Hubble to study the nature of nebulae and to a classification of nebulae based on their appearance. It also convinced him, by 1917, that the great spirals, in particular, must lie beyond the Milky Way.

The development of these ideas was delayed because as soon as Hubble had completed his Ph.D. he volunteered to serve in the US Army, following the United States's entry into the First World War in April 1917. He served in France and reached the rank of major, but never saw action. It wasn't until September 1919 that Hubble eventually joined the staff at Mount Wilson Observatory, where he was one of the first people to use the new 100-inch telescope. He also took the opportunity to develop the ideas from his Ph.D. thesis into a full classification scheme which he completed in 1923. Hubble always used the term nebulae for the objects he was describing, but he was convinced that they lay outside the Milky Way; as he was soon proved right, in line with modern usage I shall call them galaxies. The most important thing to emerge from Hubble's early work is that there are indeed different kinds of galaxy, and the giant spirals are simply the most obvious of these objects.

Apart from a relatively small number of relatively small, irregularly shaped galaxies like the Small Magellanic Cloud (and its bigger counterpart the Large Magellanic Cloud), all galaxies can be defined according to their shape. The term elliptical galaxy is used for those which appear to be anything from spherical to the shape of an elongated lens, but have no obvious internal structure. Spirals may have more tightly wound or more open spiral structures, and in all cases there are examples in which the spiral arms start at the centre of the galaxy, and examples in which the spiral arms seem to be connected to the ends of a bar of stars across the centre of the galaxy. Hubble thought that there was an evolutionary sequence in which an open spiral of either kind gradually became more and more tightly wound, as a result of rotation, and ended up as an elliptical. He was completely wrong, but this does not affect his classification scheme based on the appearance of galaxies. We now know that the largest galaxies in the Universe are giant ellipticals, but some ellipticals are smaller than some spirals. We also know that some of the galaxies originally regarded as 'spirals' are disc-shaped systems of stars with no discernible spiral arms at all! For this reason, it is better to use the term 'disc galaxy', which includes the ones with spiral arms; but even today many astronomers refer to 'spirals' when they are talking about essentially featureless disc galaxies.

Hubble's career at Mount Wilson overlapped briefly with that of Harlow Shapley, who left to take up a post at Harvard in March 1921. By the time Hubble began using the 100-inch telescope to try to prove that the nebulae he had been studying were other galaxies, the more senior astronomer was no longer around to object. With ever-improving observations, the island universe idea was, in any case, beginning to gain support in the early 1920s. A Danish astronomer, Knut Lundmark, who visited both the Lick Observatory and the Mount Wilson Observatory at that time, obtained photographic images of a nebula (galaxy) known as M33 which were good enough to convince him, although not Shapley, that the granulated appearance of the image showed that the nebula was made of stars. In 1923, several variable stars were discovered in the nebula NGC 6822, but it took a year before they could be identified as Cepheids, and by then Hubble had already made the breakthrough discovery of Cepheids in M31, the Andromeda Nebula.

He wasn't actually looking for Cepheids. With his classification scheme completed, in the autumn of 1923 Hubble followed up one of the main lines of Curtis's argument by starting a series of photographic observations with the 100-inch telescope, aimed at discovering novae in one of the spiral arms of M31. Almost immediately, in the first week of October that year, he found three bright spots of light which looked like novae on the photographic plates. Because the 100-inch telescope had been operating for several years, there was already an archive of photographs which included observations of the same part of M31, obtained by several observers, including Shapley and Milton Humason, who was to become Hubble's closest collaborator in the years that followed. These plates showed that one of the three bright spots that Hubble had tentatively identified as novae was, in fact, a Cepheid, with a period of a little more than 31 days. Using Shapley's calibration of the Cepheid distance scale, this immediately gave a distance of nearly a million light years (300 kpc), three times bigger than even Shapley's estimate of the size of the Milky Way Galaxy. The whole distance scale was later revised, partly because of the problems caused by interstellar extinction, and we now know that M31 is actually about 700 kpc away, roughly equivalent to about 20 times the diameter of the Milky Way. But what mattered in 1923 was that at a stroke, with almost his first observations of the nebula, Hubble had shown that it is indeed a galaxy more or less like our own, located far outside the Milky Way.

Over the following months, Hubble found one more Cepheid and nine novae in M31, all giving him roughly the same distance estimate, and other Cepheids and novae in other nebulae. He put everything together in a paper presented to a joint meeting of the American Astronomical Society and the American Association for the Advancement of Science held in Washington, DC, on 1 January 1925. Hubble was not present at the meeting, where the paper was read on his behalf by Henry Norris Russell. But there was no need for personal advocacy. The consensus of the meeting was that the nature of the nebulae had at last been determined, and that the Milky Way Galaxy is just one island in a much bigger Universe. Even before that meeting, Hubble had written to Shapley to tell him about the discoveries. The astronomer Cecilia Payne-Gaposchkin, who had started her Ph.D. studies under Shapley's supervision in 1923, happened to be in his office when he read the letter. 'Here', he said as he offered it to her, 'is the letter that destroyed my universe.' The Great Debate was over. It may have been some consolation for Shapley that Hubble's successful

3. The dome of the 100-inch Hooker Telescope on Mount Wilson, used by Edwin Hubble to measure distances to galaxies

use of the Cepheid technique lent weight to Shapley's model of the Milky Way, and in particular to the displacement of the Sun from the centre of our Galaxy.

But if Shapley's universe had been destroyed, what was the new universe – Hubble's universe – like? The Universe is so big that even with the 100-inch telescope Hubble was only able to obtain images of Cepheids in what turned out to be very nearby galaxies. Observers working with lesser telescopes were even more handicapped. Fascinated – almost obsessed – with the idea of measuring the scale of the Universe, Hubble had to find other ways to measure the distances to galaxies beyond the range of the Cepheid technique, and he set about this task in the middle of the 1920s.

Hubble put together a series of stepping stones which observers could use to reach farther and farther out into the Universe. Cepheids are just bright enough to give the distances to a few of the nearest galaxies, only a few dozen before the advent of the space telescope named after Hubble and launched in 1990, but novae are a little brighter than Cepheids and can be seen farther away. Once the distance to M31 had been determined from Cepheids, Hubble was able to use this to calibrate the brightness of novae seen in that galaxy, and then, making the assumption that all novae have the same intrinsic brightness, to use observations of novae to measure distances to galaxies that are a little farther away. With the ability of the 100-inch and its successors to resolve individual stars in nearby galaxies, other techniques became feasible. The brightest individual stars in galaxies are also much brighter than Cepheids, and could be used as distance indicators in the same way, this time making the assumption that the brightest stars in any galaxy will be about as bright as the brightest stars in any other galaxy, since there must be some upper limit to how bright a star can be. He could also identify globular clusters in other galaxies, and guess that the brightest globulars in each galaxy must have roughly the same intrinsic brightness as each other. Supernovae, once they were understood, were later added to the chain in the same way.

More rough and ready estimates were based on the brightnesses of whole galaxies, and on their apparent (angular) sizes on the sky. If every spiral galaxy was exactly as bright as M31 and each the same size as M31, it would be easy to measure their distances by comparing their observed properties with the properties of M31. Unfortunately, this is far from being the case, and Hubble knew it; but for want of anything better, he tried to compare the observed properties of galaxies that looked much the same as one another in order to get at least some guide to their distances.

None of these techniques is perfect, but wherever possible Hubble applied as many of the techniques as he could to each individual galaxy, hoping that the errors and uncertainties would average out. This all took time, but in 1926 Hubble had begun to build up a picture of the distribution of galaxies around the Milky Way Galaxy. There was just enough data for him to contemplate taking

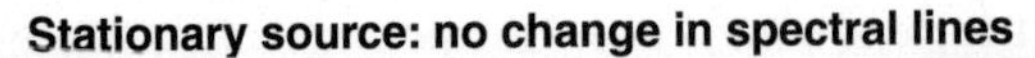

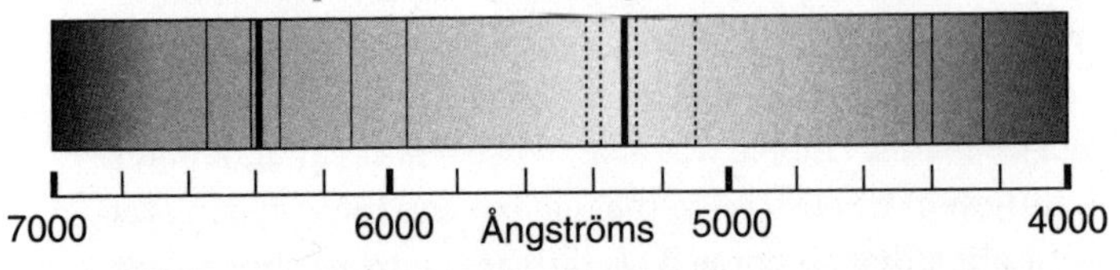

Approaching source: spectral lines shift towards the blue end of the spectrum

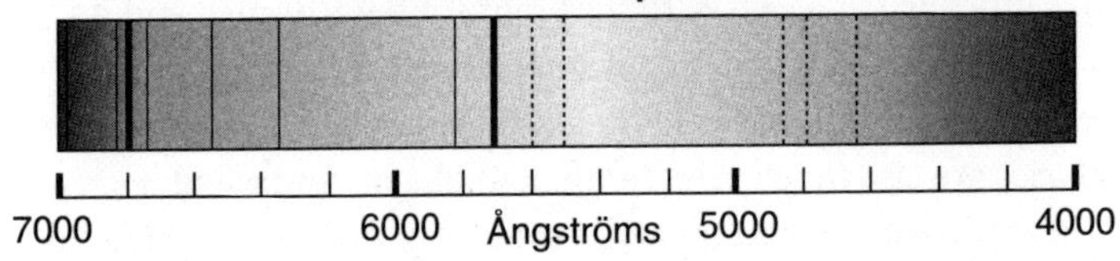

Receding source: spectral lines shift towards the red end of the spectrum

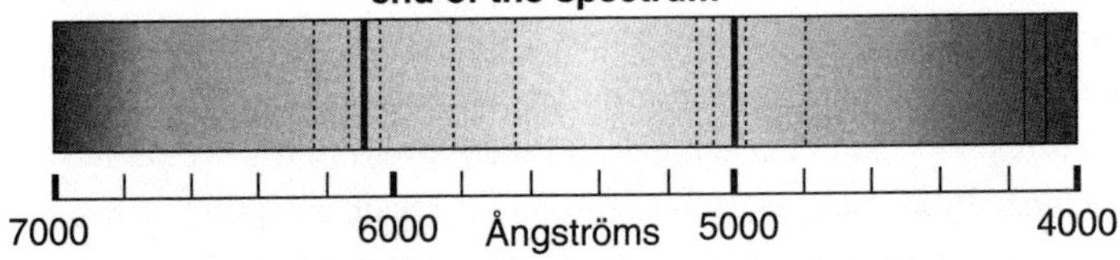

4. The velocity and direction of travel of the star in relation to the observer determine the amount by which the banding in the spectrum shifts. When a radiating body is moving away from the observer, the waves emitted become 'stretched', the wavelength lengthens, and the spectral lines shift towards the red end of the spectrum. If the body is approaching, the wavelength is compressed, and the lines shift towards the blue end of the spectrum. Redshift can be used to calculate an object's recession velocity

a great leap outward by following up a hint already present in the redshift data obtained by Vesto Slipher and a few other people.

By 1925, analyses of the light from what were now known to be other galaxies had revealed 39 redshifts and still the same two blueshifts. In fact, Slipher had been the first person to measure all but four of these redshifts, but he soon reached the limit of what was possible with the telescope he was using at the Lowell Observatory, a 24-inch (60-cm) refractor, ending up with 43 redshifts. There was a hint – barely – from these data that larger redshifts were associated with more distant galaxies. Several people had noticed this, but Hubble, by now an established astronomer with access to the best telescope in the world, was the man in the right place at the right time to try to prove that this was the case. His motivation was to find out whether there was a precise relationship between redshifts and distances that he could use as the final step in his chain, making it possible to measure distances across the Universe simply by measuring redshifts.

In 1926 Hubble deliberately set out to find a link between redshifts and distances for galaxies. He already had many distances, and would determine more over the years that followed, but the 100-inch had never been used for redshift work, and he needed a colleague able and willing to set the telescope up for this difficult work, and then carry out the painstaking measurements. He chose Milton Humason, a superb observer but clearly junior to Hubble, so that it would be obvious to the outside world who was the team leader. After all the hard work of adapting the 100-inch to its new role, Humason deliberately chose for his first redshift measurement a galaxy too faint to have been studied in this way by Slipher. He obtained a redshift corresponding to a velocity of about 3,000 km per second, more than twice as large as any redshift measured by Slipher. The Hubble–Humason partnership was up and running.

By 1929, Hubble was convinced that he had found the relationship between redshift and distance. Not only that, it was the simplest relationship he could have hoped to find – redshift is proportional to distancc, or, putting it the way round that mattered to Hubble, distance is proportional to redshift. A galaxy with a redshift twice as large as that of another galaxy is simply twice as far away as the nearer galaxy. The first results of the collaboration, published in 1929, gave data for just 24 galaxies which had both known redshifts and known distances, from which Hubble calculated that the constant of proportionality in the redshift–distance relation was 525 km per second per Megaparsec. That is, a galaxy with a redshift corresponding to a velocity of 525 km per second would be one million parsecs (3.25 million light years) away, and so on. The choice of this particular number looked as much like wishful thinking as anything else, because the limited amount of data was not really good enough to justify the precision of the quoted number. But in 1931 Hubble and Humason together published a paper updating these results with a further 50 redshifts, going out to a distance equivalent to a velocity of 20,000 km per second, and fitting the number Hubble had obtained three years earlier much more closely. Clearly, he had already had some of these data in 1929, but had chosen, for whatever reason, not to publish them at the time.

Hubble neither knew nor cared why the redshift–distance relation existed. He didn't even claim that it meant that other galaxies are moving away from us. Although redshifts are conventionally quoted in units of km per second, there are other ways than motion through space known to produce them (for example, a strong gravitational field) and Hubble was careful to consider that processes unknown in the 1930s might be at work. In his book *The Realm of the Nebulae*, he wrote:

> Red-shifts may be expressed on a scale of velocities as a matter of convenience. They behave as velocity-shifts behave and they are very simply represented on the same familiar scale, *regardless of the ultimate interpretation*. The term 'apparent velocity' may be used in carefully considered statements, and the adjective always implied where it is omitted in general usage. (Emphasis added.)

Whatever the origin of the redshift–distance relation, it did indeed prove the ultimate tool for measuring the scale of the Universe,

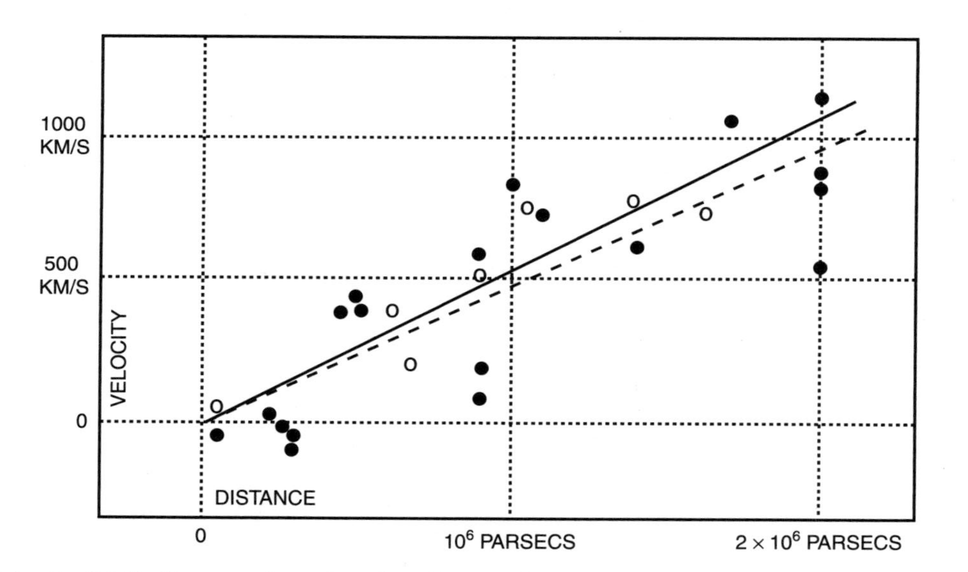

5. Hubble's original 'redshift–distance relation' was based on a rather optimistic interpretation of the data published in 1929; by 1931, his work with Humason made a much more convincing case

and the constant of proportionality became known as Hubble's Constant, or *H*. Since 1931, the aim of all other measurements of distances beyond the Milky Way has simply been to calibrate the Hubble Constant. But before looking at what all this implies for the understanding of galaxies and their place in the Universe at large, it seems appropriate to sum up the present understanding of our own home in space, the Milky Way, an ordinary spiral galaxy.

Chapter 3
Our island

Since the 1920s our understanding of the Milky Way Galaxy has increased dramatically, largely because of the continuing development of observing techniques and technology. As well as having larger and better telescopes to observe in visible light (including the Hubble Space Telescope), we have data obtained by radio telescopes, in the infrared part of the spectrum and from X-ray detectors and other instruments carried into space on satellites. Sensitive electronic detectors are able to obtain much more information about faint sources than is available from photographs or from the kind of spectroscopic instruments available to Hubble and his contemporaries, and the power of modern computers makes it much easier than it was in his day to compare theoretical predictions with observations.

The most profound discovery about the Milky Way made since the 1920s is that all of the bright stars make up only a small fraction of the total amount of mass in our Galaxy. From the way that the whole system rotates, it is clear that the bright disk is held in the gravitational grip of a roughly spherical halo of dark matter which has about seven times as much mass as everything that Hubble would have thought of as the Galaxy put together. This has profound implications for our understanding of the Universe at large, since the same ratio of ordinary matter to dark matter seems to apply across the Universe. These cosmological implications are discussed by Peter Coles in *Cosmology: A Very Short Introduction*. The most important point, apart from the existence of the dark matter, is that it is not simply cold gas or dust. It is not made of the kind of particles – atoms and so on – that the Sun and stars, and ourselves are made of, but is something else entirely. Since nobody knows exactly what it is, it is simply referred to as Cold Dark Matter, or CDM.

Our Sun is a typical star. Some contain more mass, some less, but they all work in the same way, converting light elements (in particular hydrogen) into heavier elements (in particular, helium) in their interiors by nuclear fusion, releasing the energy that keeps the stars shining. Overall, it is estimated that there are several (at least three) hundred billion stars in the Milky Way, spread across a disc about 28 kiloparsecs (just over 90 thousand light years) across. There is some uncertainty about the exact size (it is difficult to measure the size of a forest from inside it), so these numbers are often rounded off to 30 kpc and 100,000 light years. There is a bulge of stars in the centre of the disc, which would give it the appearance, if viewed edge-on from outside, of two fried eggs stuck back to back. The whole disk is surrounded by the spherical halo of old stars and globular clusters, which contain the oldest stars in the Galaxy. Nearly 150 globular clusters are known, and there must be another 50 or so that we cannot see because the bright band of light of the Milky Way is between us and them.

Astronomers can study the way stars move through space using the Doppler effect. This shifts the lines in the spectrum of a star towards the red end of the spectrum if it is moving away, and towards the blue if it is moving towards us. The size of the effect reveals the star's velocity. This is exactly equivalent to the way the sound emitted from a moving object – for example, the siren on an ambulance – is deepened if the vehicle is moving away but raised in pitch if it is moving towards you. Christian Doppler predicted the effect in 1842, then measured it using trumpeters blowing a steady note as they moved past on a train. Superficially, the

6. The starburst galaxy M82. This is a composite image combining data from WFPC2 and the 3.5-metre telescope on Kitt Peak in the United States

effect resembles the redshift seen in the light of galaxies; but the cosmological redshift is not caused by motion through space and is not a Doppler effect.

The Sun is about two-thirds of the way out (a bit less than 10 kpc) from the centre of the Milky Way to the edge of the visible disc. Like other stars in the disc, it moves around the centre of the Galaxy, at a speed of about 250 km per second in a roughly circular orbit, and takes a little less than 250 million years to complete one circuit. The ages of stars can be determined by comparing their overall appearance (especially their colour and brightness) with theoretical models of how they change as they consume their nuclear fuel; in the case of the Sun this is confirmed by using measurements of the radioactivity in rocks and meteorites to infer the age of the Solar System. The Sun and Solar System have been around for about 4.5 billion years, long enough to have completed about 20 orbits of the galactic centre. Since the first humans, modern *Homo sapiens*, appeared on Earth, the Solar System has completed just under one-thousandth of its present circuit. The oldest stars are a little over 13 billion years old, three times the age of the Sun.

Outside the central bulge, the disc of the Galaxy is only about 300 kpc (roughly 1,000 light years) thick. The Solar System is only 6 or 7 parsecs above the centre of the plane of the disc. Viewed from above, the Galaxy's resemblance to a fried egg would be spoiled by the bar, 8 or 9 kpc long, across the middle of the bulge, but it would be possible to pick out four quite tightly wound spiral arms twining outward from the centre. As in other disk galaxies, the spiral arms are bright because they contain many hot young stars in the first flush of youth. These stars are big as well as bright. The bigger (more massive) a star is, the more intensely it has to burn its nuclear fuel to hold itself up against the pull of gravity, and the more quickly it uses up its fuel. Spiral arms are the sites of star formation. Smaller, long-lived stars, like the Sun, are also formed in spiral arms, but do not burn so brightly. The Solar System is at present in a lesser spur of stars known as the Orion Arm, or simply the Local Arm, which forms a kind of bridge between two of the major arms. Shapley was right in thinking that we are in a large local concentration of stars.

The young stars that are found chiefly in the spiral arms and the plane of the Milky Way (and in the discs of other galaxies) are known as Population I. The Sun is a Population I star. They contain recycled material from previous generations of stars,

7. Star-forming region in Orion, imaged in the infrared by the Spitzer Space Telescope

including the heavy elements from which planets like the Earth are composed. Older stars which are found in the halo of the Galaxy, in globular clusters and in the bulge, are known as Population II. These old stars tend to be redder than Population I stars. They formed long ago when the Galaxy was young, and are chiefly composed of the primordial hydrogen and helium that emerged from the Big Bang in which the Universe was born. The heavier elements in Population I stars and in ourselves were made inside previous generations of stars. Elliptical galaxies are largely made up of Population II stars.

If the spiral pattern seen in a galaxy like the Milky Way was not maintained in some way, it would soon get smeared out, within about a billion years, as the stars moved around the Galaxy in their orbits. It persists because it is a wave of star formation maintained by clouds of gas and dust moving around the Galaxy in their own orbits and being squeezed as they cross the spiral arms. The young stars are simply the most visible feature of a shock wave travelling around the Galaxy, similar to the shock wave of a sonic boom.

An analogy that is often made is with the kind of traffic jam that occurs on a busy motorway when there is a large, slow-moving vehicle occupying the inside lane. As the faster traffic comes up behind the large vehicle it is squeezed into the outer lanes and makes a moving traffic jam which disperses on the other side of the large vehicle. The traffic jam moves along the motorway at a steady speed, but it is constantly changing as new cars join the back and others leave the front. In the same way, a spiral arm moves around the Galaxy at a constant speed, but new clouds of gas and dust are constantly joining it, being squeezed, and then going on their way. Some of these clouds get squeezed sufficiently to trigger star formation, in a self-sustaining process.

But although it is self-sustaining, this is not a very efficient process. If it were, then by now the Milky Way would have formed all of the gas and dust it contains into stars. In fact, only a few times as much material as there is in the Sun (a few solar masses of material) is converted into new stars each year in our Galaxy. This roughly balances the amount of material thrown back out into space by old stars when they die, so the processes of star birth, life, and death are able to continue for many billions of years in a disc galaxy. This also implies that very many stars must have been born in a short space of time when the Milky Way formed, before it settled down. Such spectacular events, known as starbursts, are indeed seen in other galaxies.

It is difficult for a cloud of gas and dust to collapse to form a star (or several stars) for two reasons. First, all clouds are rotating, if only slightly, and as they contract they will spin faster, resisting the pull of gravity. They have to break up in such a way that their angular momentum is dissipated in some way. Second, a collapsing cloud will get hot, as gravitational energy is released, and unless this heat can be radiated away it will prevent any further collapse. The angular momentum problem is solved by clouds breaking up into several stars, so that the angular momentum of the cloud is converted into the angular momentum of the stars orbiting one another. On average, out of every 100 newly born star systems, 60 are binaries and 40 are triples. Solitary stars like the Sun are later ejected from triple systems formed in this way. The heat problem is solved because the clouds contain molecules such as carbon monoxide, which warm up and radiate the heat away in the infrared part of the spectrum. But star formation is still a difficult process – the wonder is that there are any stars at all.

Star formation begins in large complexes of gas, perhaps a thousand parsecs across and containing ten million solar masses of material, within which an individual cloud may be a few tens of parsecs across and contain a few hundred thousand solar masses of material. The initial squeeze to cause the collapse of a cloud most probably comes from the explosion of a massive star, a supernova. Turbulence within the collapsing cloud leads to the formation of cores about a fifth of a light year across, containing about 70 per cent as much mass as our Sun. But only a few per cent of the mass of the whole cloud gets converted into cores in this way. When a star forms, it starts out as an even smaller inner core, with only one thousandth of the mass of the Sun, reaching the density necessary to turn itself into a star. The rest of the mass of the star is added as material from the surrounding cloud close enough to be pulled in by gravity falls on to the core, so the eventual mass of the star depends on how much material there is nearby. Once the stars begin to shine, the radiation from them blows away the rest of the surrounding material.

The whole process is over very quickly. A cloud collapses to make stars and the hot young stars blow away the leftover material to leave behind a cluster of stars, all within about ten million years. The late stages of this process can be seen in the nearby Orion Nebula. But some of the young stars in some clusters will be much more massive than the Sun, and will use up their nuclear fuel very quickly. These are the stars which end their lives by exploding as supernovae, sending shock waves out through the interstellar material and triggering the collapse of other clouds of gas and dust. This seems to be a self-sustaining process which keeps a galaxy like the Milky Way in a steady state with the aid of negative feedback. If a larger than average number of stars form in one generation or one location, the energy from them will disperse the gas and dust over a wide region, reducing the number of stars in the next generation; but if only a few stars form, there will be plenty of gas and dust left over to make more stars next time the cloud is squeezed. The natural tendency is for the process to shift back towards the average. And because the kind of stars that form supernovae burn out in only a few million years (compare that with the 4.5 *billion* years that the Sun has been around so far) all of this activity takes place within the vicinity of the spiral arms, helping to maintain the spiral pattern.

The central region of our Galaxy, around which the whole spiral pattern rotates, is more than just the mathematical centre of the disc. There is a black hole containing 2.5 million times as much mass as our Sun at the centre of the Milky Way, and as we shall see in Chapter 7, such black holes hold the key to the existence of galaxies.

Most popular accounts of black holes concentrate on much smaller objects, with masses only a few times that of our Sun. Such objects form if a star at the end of its life has more than about three times the mass of the Sun today. Such a stellar cinder, no longer generating heat in its interior because all of its fuel is exhausted, cannot hold itself up under its own weight, and collapses, shrinking (according to the general theory of relativity) into a point of zero volume, called a singularity. Atoms and the particles they are made of, protons, neutrons, and electrons, are crushed out of existence in the process. Almost certainly, the general theory of relativity breaks down before the singularity is reached, but long before that happens the gravitational attraction of the collapsing object becomes so powerful that nothing can escape, not even light. This is where black holes get their name. One way of thinking about what is going on is that the escape velocity from a black hole exceeds the speed of light. Since nothing can travel faster than light, nothing can escape from a black hole.

In fact, any object will become a black hole if it is sufficiently compressed. For any mass, there is a critical radius, called the Schwarzschild radius, for which this occurs. For the Sun, the Schwarzschild radius is just under 3 km; for the Earth, it is just under 1 cm. In either case, if the entire mass of the object were squeezed within the appropriate Schwarzschild radius it would become a black hole.

But although black holes themselves are invisible, they exert a gravitational influence on their surroundings, and this can lead to violent and easily detectable activity in their vicinity. We know that stellar mass black holes exist because some of them are in orbit around ordinary stars, forming binary systems. The direct effect of the gravity of the black hole on the binary orbit of its partner reveals the mass of the black hole, and matter pulled off from the companion streams down towards the black hole, funnelling into its 'throat'. There, the infalling matter gets hot enough to emit X-rays as the particles in the stream speed up and collide with one another.

All these black holes are associated with matter squeezed to very high densities. The black hole at the centre of the Galaxy is a different kind of beast. Curiously, though, such supermassive black holes were the first to pique the curiosity of theorists, long before Albert Einstein came up with the general theory of relativity. In 1783, John Michell, a Fellow of the Royal Society, pointed out that, according to Newton's theory of gravity, an object with a diameter 500 times that of the Sun (about as big across as the Solar System) but with the same density as the Sun would have an escape velocity greater than that of light. (Michell did not use the term 'escape velocity' but in modern language that is what he was talking about; Einstein's theory, of course, makes the same prediction.) This need not involve superdensities at all, since the overall density of the Sun is only about one-and-a-half times the density of water. The Frenchman Pierre Laplace reached the same conclusion independently in 1796, and commented that, although these dark objects could never be seen directly, 'if any other luminiferous bodies should happen to revolve about them we might still perhaps from the motions of these revolving bodies infer the existence of the central ones'. Two centuries later, that is exactly how the black hole at the heart of the Milky Way was discovered.

The centre of the Milky Way lies in the same direction on the sky as the constellation Sagittarius, but much farther away. The constellations, named in ancient times, are patterns of nearby stars, which look bright simply because they are close to us. The names are still used by astronomers to indicate which part of the sky – which direction – an object lies in. That is why M31 is also known as the Andromeda Nebula (or Andromeda Galaxy), even though it is a couple of million light years farther away than the stars in the constellation Andromeda, and has nothing at all to do with them. In the same way, a powerful source of radio noise at the centre of our Galaxy is known as Sagittarius A, even though it has nothing to do with the stars of the constellation Sagittarius.

It only became possible to study the centre of our Galaxy when radio telescopes and other instruments that do not rely on visible light became available. There is a great deal of dust in the plane of the Milky Way, responsible for the interstellar extinction that plagued early attempts at determining the distance scale and providing some of the raw material for new generations of stars. This blocks out visible light. But longer wavelengths penetrate the dust more easily. That is why sunsets are red – short wavelength (blue) light is scattered out of the line of sight by dust in the atmosphere, while the longer wavelength red light gets through to your eyes. So our understanding of the galactic centre is largely based on infrared and radio observations.

More detailed studies showed that Sagittarius A is actually made up of three components lying close to one another. One is the expanding bubble of gas associated with a supernova remnant, one is a region of hot, ionized hydrogen gas, and the third, dubbed Sgr A*, is at the very centre of the Galaxy.

There is certainly plenty of activity around Sgr A*. Infrared studies reveal a dense cluster of stars in which 20 million stars like the Sun are packed into a volume one parsec across, where the stars are on average only a thousand times farther apart than the distance from the Earth to the Sun, and collisions occur every million years or so. There is a massive ring of gas and dust around this cluster, extending out from about 1.5 pc to a distance of 8 pc (some 25 light years), with traces of shock waves from recent explosive events, and both X-rays and even more energetic gamma rays pour out of the central region.

But for all of this high-tech stuff, the best evidence for the presence of the black hole comes from the kind of study Laplace envisaged. Observations made at infrared wavelengths using a telescope with a 10-metre diameter mirror at the Mauna Kea Observatory in Hawaii provided measurements of the speed with which 20 stars close to the galactic centre are moving. The stars are orbiting the galactic centre at speeds of up to 9,000 km per second, which converts to nearly 30 million miles per hour. They are moving so fast that even though they are so far away – nearly 10 kpc – their positions are seen to change in photographs taken at intervals a few months apart over a few years, and by putting such pictures together it is possible to make a movie which actually shows the orbits of the innermost of these stars. The orbital motion tells us that the stars are in the grip of an object with between two and three million times the mass of our Sun. Since this is contained in a volume of space no bigger across than the radius of the Earth's orbit around the Sun, it is definitely a supermassive black hole.

The black hole is relatively quiet today, because it has swallowed up all of the matter in its immediate locality. The activity we can detect now results from a dribble of matter falling into the hole from the surrounding ring of stuff; all it needs to 'eat' each year in order to maintain the present level of activity is a mass equivalent to about 1 per cent of the mass of our Sun, releasing gravitational energy as the matter falls into the hole. Things must have been different long ago when the Galaxy was young and the region around the black hole had not been swept clear of gas and dust; I shall discuss this later, but it is clear that supermassive black holes are the seeds from which galaxies grew.

The way stars move farther out from the galactic centre can also tell us something about the way the Galaxy got to be the way it is today. The orderly structure described so far, with bulge, disc, and halo components, is not quite the whole story. When astronomers look in detail at the compositions of individual stars and the way they are moving, they find that against the background of the many stars that move together in the Milky Way they can pick out long, thin streams of stars which have similar makeup to one another, different from that of the background stars, and are moving in the same direction as each other, at an angle to the motion of most of the stars in that part of the sky.

Nine or ten such streams have now been identified (the exact number depends on how reliable you think the evidence is), with more still being found. They range in mass from a few thousand to a hundred million solar masses of material and in length from 20,000 to a million light years. Very often, these star systems can be traced as tenuous connections to a globular cluster, or to one of the 20 or so small galaxies that orbit the Milky Way Galaxy like moons orbiting around a planet. The most spectacular of these star trails, from our perspective, is called the Sagittarius stream. It extends over a curving span of more than a million light years, and joins the Milky Way to the so-called Sagittarius dwarf elliptical. Another stream, seen in the direction of the constellation Virgo and therefore called the Virgo stellar stream, is moving almost perpendicular to the plane of the Milky Way, and is associated with another dwarf galaxy.

This kind of evidence explains the origin of the star streams. Small galaxies that come too close to our own Galaxy get broken up and dissipated by the gravitational forces – tides – they encounter, trailing a stream of stars as they move in their orbits around the Milky Way. The Sagittarius dwarf is in the final stages of this process, barely discernible today as a coherent group of stars. Eventually, there will be nothing left but the star stream, which will merge with the Milky Way and finally lose its identity.

This is a clear indication that the Milky Way reached its present size through a kind of intergalactic cannibalism, swallowing up its lesser neighbours. Using powerful statistical techniques, astronomers are even able to work backwards from the observations of how star streams are moving today to reconstruct the ghosts of former satellite galaxies, like palaeontologists reconstructing the appearance of a dinosaur from a few fossil remains. And as the icing on the cake, the shape of the orbits of these star streams tells us that the extended halo of dark matter in which the Milky Way is embedded is spherical, not ellipsoidal.

These galactic interactions are not, though, confined to occasions when a large galaxy swallows up its small neighbours. As Vesto Slipher discovered, light from the Andromeda galaxy shows a blueshift corresponding to an approaching velocity of more than 100 km per second (approaching 250,000 miles per hour). The reason it does not show a redshift is because, as Hubble realized, the cosmological redshift is not caused by motion through space. At the distance to the Andromeda galaxy, the cosmological redshift would be tiny, equivalent in velocity terms to less than half the Andromeda galaxy's observed blueshift. But galaxies do move through space, and these motions cause Doppler effects which are superimposed on their cosmological redshifts.

For all but the nearest of our neighbours, the cosmological redshift is much bigger than any Doppler effect, and dominates. But in the case of the Andromeda galaxy, the Doppler effect is much bigger than the cosmological redshift. The Andromeda galaxy really is moving rapidly towards us, and will collide with the Milky Way in about four billion years from now – coincidentally, just when the Sun is nearing the end of its life. Such a collision between roughly comparable disc galaxies will lead to a merger. The stars in each galaxy are separated by such great distances that there will be no collisions between stars in the two discs, but computer simulations show that the gravitational forces will cause the structure of the two discs to be destroyed as the stars merge into one system, forming a giant elliptical galaxy.

All of the discoveries described in this chapter would be important if they only told us about the Milky Way, our island home. But they are doubly important because there is strong evidence that the Milky Way Galaxy is a completely ordinary disc galaxy, a typical representative of its class. Since that is the case, it means that we can confidently use our inside knowledge of the structure and evolution of our own Galaxy, based on close-up observations, to help our understanding of the origin and nature of disc galaxies in general. We do not occupy a special place in the Universe; but this was only finally established at the end of the 20th century.

Chapter 4
Interlude: galactic mediocrity

It can be argued that the scientific revolution began in 1643, with the publication by Nicolaus Copernicus of a book, *De Revolutionibus Orbium Coelestium*, setting out the evidence that the Earth is not at the centre of the Universe, but moves around the Sun. Since then, it has become appreciated that the Sun is just an ordinary star, occupying no special place in our Milky Way Galaxy, let alone in the Universe, and that humankind is just one species of life on Earth, occupying no special place except from our own parochial point of view. With tongue only slightly in cheek, some astronomers say that all of this is evidence in support of the 'principle of terrestrial mediocrity', which says that our surroundings are completely lacking in any special features, as far as the Universe is concerned. This might be a humbling thought for anyone still harbouring pre-Copernican ideas; but if it is correct, it does mean that we can extrapolate from our observations of our surroundings to draw meaningful conclusions about the nature of the Universe at large. If the Milky Way is mediocre, then billions of other galaxies must be very much like the Milky Way, just as one city suburb looks much the same as another city suburb.

But in the decades following Hubble's first measurements of the cosmological distance scale, the Milky Way still seemed like a special place. Hubble's calculation of the distance scale implied that other galaxies are relatively close to our Galaxy, and so they would not have to be very big to appear as large as they do on the sky; the Milky Way seemed to be by far the largest galaxy in the Universe. We now know that Hubble was wrong. Because of the difficulties he struggled with, including extinction and a serious confusion between Cepheids and other kinds of variable stars, the value he initially found for the Hubble Constant was about seven times bigger than the value accepted today. In other words, all the extragalactic distances Hubble inferred were seven times too small. But this was not realized overnight. The cosmological distance scale was only revised slowly, over many decades, as observations improved and one error after another was corrected. I do not intend to take you through all the steps here, but to present the simplest and most direct evidence, using the latest and best observations, for the galactic mediocrity of the Milky Way.

Even in the 1930s, some scientists were unhappy about the idea that the Milky Way might be an unusually large galaxy. The person who felt most strongly about this, and expressed his doubts most forcefully, was the astronomer Arthur Eddington, best remembered as the leader of the eclipse expedition of 1919 that verified the predictions of Einstein's general theory of relativity. Eddington was a firm believer in what is now know as the principle of terrestrial mediocrity, and in his book *The Expanding Universe*, published in 1933, he wrote:

> The lesson of humility has so often been brought home to us in astronomy that we almost automatically adopt the view that our own galaxy is not specially distinguished – not more important in the scheme of nature than the millions of other island galaxies. But astronomical observation scarcely seems to bear this out. According to the present measurements the spiral nebulae, though bearing a general resemblance to our Milky Way system, are distinctly smaller. It has been said that if the spiral nebulae are islands, our own galaxy is a continent. I suppose that my humility has become a middle-class pride, for I rather dislike the imputation

> that we belong to the aristocracy of the universe. The Earth is a middle-class planet, not a giant like Jupiter, nor yet one of the smaller vermin like the minor planets. The sun is a middling sort of star, not a giant like Capella but well above the lowest classes. So it seems wrong that we should happen to belong to an altogether exceptional galaxy. Frankly I do not believe it; it would be too much of a coincidence. I think that this relation of the Milky Way to other galaxies is a subject on which more light will be thrown by further observational research, and that ultimately we shall find that there are many galaxies of a size equal to and surpassing our own.

Eddington's argument made complete sense, and it eventually turned out that he was right. But in 1933, this was based on no more than his 'middle-class pride'. After all, some galaxies *are* bigger than others, and if the universe really was dominated by one huge galaxy surrounded by a host of smaller ones you could argue that it was more likely than not that we should find ourselves on the continent rather than on one of the islands. The only way to settle the issue would be to have accurate distance measurements to a large enough number of other disc galaxies to have a good understanding of their sizes in relation to that of the Milky Way. That meant Cepheid distances, and enough of these simply were not available before the launch of the Hubble Space Telescope in 1990 and its repair in 1993.

The importance of determining the cosmological distance scale accurately, more than half a century after Hubble's pioneering work, was still so great that it was a primary justification for the existence of the Hubble Space Telescope (HST). The expressed aim of the Hubble key project was to use the telescope to obtain data from Cepheids in at least 20 galaxies and use them to pin down the value for the Hubble Constant to an accuracy of plus or minus 10 per cent. By the end of the observing phase of the key project, distances to 24 galaxies had been determined accurately using Cepheids. While the Hubble team moved on to the next phase, using these data to calibrate other indicators

8. The Hubble Space Telescope in orbit

such as supernovae, the basic Cepheid data were made available to other astronomers. Together with Simon Goodwin and Martin Hendry, at the University of Sussex, in 1996 I used these Cepheid distances, the 'further observational research' Eddington had called for, to test his belief that the Milky Way is just an ordinary spiral. (The results were published in 1998.)

Using mostly HST data and some from ground-based telescopes, we found that there were 17 spirals, closely resembling the Milky Way in appearance, which had well-determined distances. The standard way to measure the angular diameter of a galaxy is, in effect, to draw contour lines of brightness (isophotes) around it, and to make the cut-off at a certain brightness level. With angular diameters determined in this way and accurate distances from the Cepheids, the true linear sizes of the 17 galaxies followed.

The hardest part of the project turned out to be measuring an equivalent diameter for the Milky Way – the classic problem of the difficulty of seeing the wood for the trees. But observations of the distribution of stars within the Milky Way made it possible to work out what it would look like from above, and this gave us an equivalent isophotal diameter of just under 27 kiloparsecs. The big question was, how would this compare with the diameters of the other 17 galaxies? The short answer is that the average diameter of all the 18 galaxies in our sample, including the Milky Way, was just over 28 kpc. Exactly as Eddington had surmised, the Milky Way is an ordinary spiral, with a diameter fractionally, but not significantly, smaller than the average. Most definitely, it is *not* a continent among islands. But nor is it significantly smaller than average. The Milky Way is, in a word, mediocre.

Among other things, this makes it possible to use observations of galaxy diameters to determine the value of the Hubble Constant, and to do so within the 10 per cent accuracy set as a target by the Hubble key project. By putting this in a cosmological context, as I shall do in the next chapter, it reveals the age of the Universe itself – the time that has elapsed since the Big Bang.

Chapter 5
The expanding universe

Modern cosmology began with Hubble's two great discoveries about galaxies – that they are other islands in space outside the Milky Way, and that there is a relationship between the redshift in the light from a distant galaxy and its distance. Together, these two discoveries mean that galaxies can be used as test particles to reveal the overall behaviour of the Universe. In particular, they show that the Universe is expanding.

Although the discovery of the redshift–distance relationship came as a surprise at the end of the 1920s, it was almost immediately realized that a mathematical theory describing this kind of universal behaviour had already been found – Albert Einstein's general theory of relativity. The general theory describes the relationships between space, time, matter and gravity. One of the key features of the theory is that space and time should not be thought of as separate entities, but as facets of a single four-dimensional entity known as spacetime. The idea of four-dimensional spacetime dates back to 1908, when Hermann Minkowski refined Einstein's special theory of relativity, which he had published in 1905. 'Henceforth,' Minkowski said, 'space by itself, and time by itself, are doomed to fade into mere shadows, and only a kind of union of the two will preserve an independent reality.'

The limitation of the special theory (the reason why it is 'special', as in a special case of something more general) is that it does not deal with gravity or with acceleration. It describes precisely the relationships between all moving objects and light (used as a general term for all electromagnetic radiation) as long as they are moving in straight lines at constant speed, and how the world would look from the point of view of any of those objects. These was a far greater achievement than such a quick summary suggests, because Einstein had in effect modified Isaac Newton's understanding of dynamics to take account of James Clerk Maxwell's understanding of light. But it was only intended as an interim step on the road to a complete theory which included gravity and acceleration as well.

Einstein achieved that with the general theory, which he completed in 1915. The simplest way to understand the general theory is in terms of Minkowski's four-dimensional spacetime. Einstein discovered that spacetime is elastic, so it is distorted by the presence of matter. Objects moving through spacetime follow curved paths around the distortions caused by the presence of matter, rather like the way a marble rolled across a trampoline will follow a curved path around the indentation made by a heavy object, such as a bowling ball, placed on the trampoline. The effect we call gravity is a consequence of the curvature of spacetime. In a famous aphorism, 'matter tells spacetime how to bend, spacetime tells matter how to move'.

Crucially, light rays also follow the appropriate curved paths through spacetime in the presence of matter. The effect is very small, unless the amount of matter involved is large, or it is squeezed into a small volume at very high density, or both. But it is just detectable in the region of space near the Sun. The general theory predicted that light from distant stars passing close to the edge of the Sun would be bent by a certain amount because of the way the Sun's mass distorts spacetime in its vicinity. From Earth, the effect would be to shift the apparent positions of the

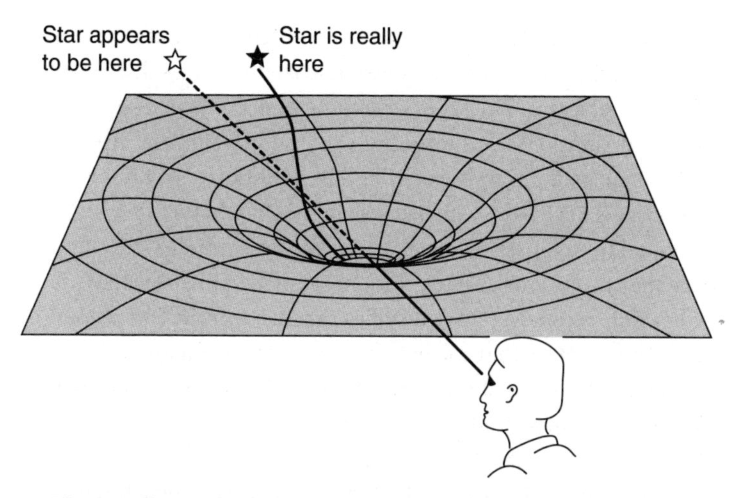

9. The Sun distorts spacetime in its vicinity, like the dent made by heavy object placed on a trampoline. Light from a distant star follows the curve in space, so the star appears to be shifted from its position when the Sun is not in the line of sight

background stars, compared with observations of the same part of the sky made when the Sun was not in the way. Since the background stars cannot be seen against the glare of the Sun, the only way to observe these changes would be during a total solar eclipse, when the Sun's light is blocked by the Moon. By great good fortune for astronomers, a suitable eclipse occurred in 1919. This was the occasion when a team led by Arthur Eddington measured the effect and found that it exactly matched the predictions of Einstein's theory; it was from that moment that Einstein became a famous celebrity, although many people never quite knew what he was famous for. Since then, the general theory has passed every test that has been devised, most recently a subtle experiment flown into space to monitor the effects of the Earth's gravity on weightless gyroscopes.

The general theory of relativity is the best theory we have to describe the overall behaviour of space, time and matter. As Einstein realized from the outset, this means that it automatically provides a description of the Universe, which is the sum total of all the space, time and matter. The trouble is, it provides descriptions of many universes. The set of equations that Einstein discovered has many solutions, as is often the case in mathematics. We are all familiar with a simple example. The equation $x^2 = 4$ has two solutions, $x = 2$ and $x = -2$, because both (2×2) and (-2×-2) are equal to 4. Einstein's equations are more complicated, and have many solutions. Some solutions describe universes that are expanding, some describe universes that are contracting, some describe universes that oscillate between expansion and collapse, and so on. But none of them, Einstein discovered to his surprise, describe a universe that is essentially still.

He was surprised because in 1917, when he worked out these solutions after completing the general theory, everyone thought that the Universe was static. The Milky Way was still thought by most astronomers to be essentially the entire universe, and although stars move around within the Milky Way, overall it is neither expanding nor contracting. The only way Einstein could obtain a mathematical description of a static universe within the framework of the general theory of relativity was to introduce an extra term into the equations, now known as the cosmological constant and usually represented by the Greek letter lambda (Λ). A dozen years later, when Hubble discovered the redshift–distance relationship, it turned out to match the mathematical description of an expanding universe in one of the simplest solutions to Einstein's equations, without the lambda term. Einstein described the introduction of the cosmological constant as 'the biggest blunder' of his career, and it was discarded by everyone except a few mathematicians who liked playing with equations for their own sake, whether or not they describe the real Universe.

The full implications of the discovery that the general theory of relativity provides a good description of our Universe are explained in Peter Coles's book. The key point to grasp, though, is that the expansion described by the equations is an expansion of

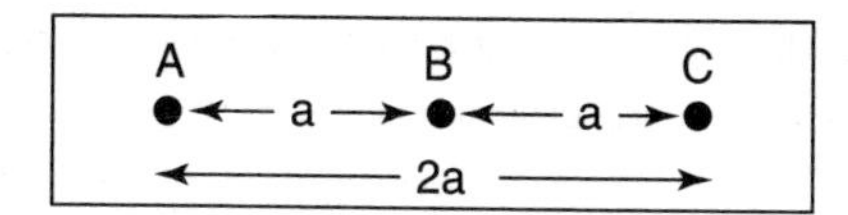

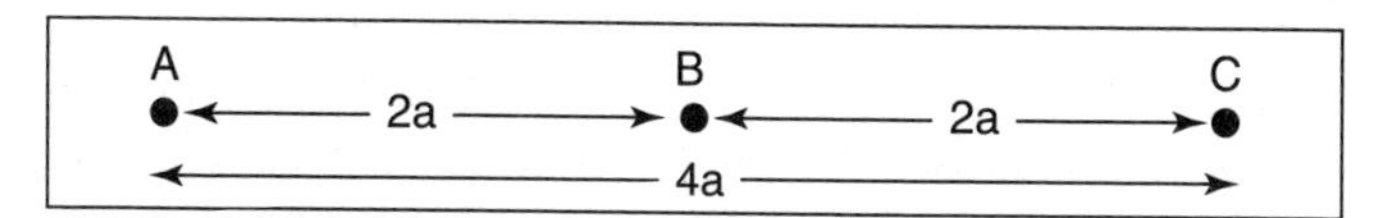

10. Expanding spacetime is like stretching a piece of rubber. The 'galaxies' A, B, and C do not move through the space between them. But when the space expands to double the distance between A and B, it also doubles the distance between every other pair of galaxies, including A and C. From the viewpoint of every galaxy in this universe, every other galaxy is receding at a rate which is proportional to its distance. Because C is twice as far away from A as B is, for example, when all distances are doubled (when the scale factor doubles) it seems that C has 'moved away' from A twice as fast as B has

space as time passes. The cosmological redshift is not a Doppler effect caused by galaxies moving outward through space, as if fleeing from the site of some great explosion, but occurs because the space between the galaxies is stretching. So the spaces between galaxies increase while light is on its way from one galaxy to another. This stretches the light waves to longer wavelengths, which means shifting them towards the red end of the spectrum.

The way the stretching occurs, though, produces redshifts which depend on relativistic effects. If we translate the redshifts into equivalent velocities, then provided the velocities involved are small compared with the speed of light, they behave in a very simple way. Redshifts are usually denoted by the letter z. If $z = 0.1$, that means an object is receding at one tenth of the speed of light (i.e. about 30,000 km per second, bigger than anything measured in the pioneering study by Hubble and Humason). A redshift of 0.2 means it is receding twice as fast, and so on – up to a point. Since nothing can travel faster than light, the largest redshift that could be produced if this simple rule held up would be 1. But when relativistic effects are taken into account, the largest possible redshift, corresponding to recession at the speed of light, is infinite. Relativistic effects become important once we are dealing with 'velocities' bigger than about a third of the speed of light. Once we take these effects into account, a redshift of 2, for example, does not mean that an object is receding from us at twice the speed of light but at 80 per cent of the speed of light; a redshift of 4 corresponds to a recession velocity of 92 per cent of the speed of light. Individual redshifts greater than 10 have now been measured, but these are very much the exception.

In fact, there are very few isolated galaxies in the Universe. Most galaxies occur in clusters, which may contain anything from a few galaxies to thousands of galaxies, held together by gravity. Individual galaxies within the cluster are moving around their mutual centre of mass, while the whole cluster is being carried along by the expansion of space. Like a swarm of bees, the galaxies move around one another while the whole swarm moves along as a unit. So when we look at the light from galaxies in a cluster, we find that there is some average redshift, which is the cosmological redshift caused by the expansion of the Universe, but that some galaxies have slightly bigger redshifts and some slightly smaller redshifts. The galaxies with smaller redshifts are the ones that are moving towards us, so that their motion *through* space contributes a Doppler blueshift which reduces the overall redshift. The galaxies with larger redshifts are the ones that are moving away from us, so that their motion *through* space contributes a Doppler redshift which enhances the overall redshift. All of is this taken account of when astronomers use the shorthand expression 'galaxies show a redshift proportional to their distance'.

The second key point about the universal expansion is that it does not have a centre. There is nothing special about the fact that we observe galaxies receding with redshifts proportional to their distances from the Milky Way. In another example of terrestrial mediocrity, whichever galaxy you happen to be sitting in, you will see the same thing – redshift proportional to distance. A simple analogy makes this clear. Imagine the surface of a perfect sphere, painted with random spots of colour to represent galaxies. If you inflate the sphere, the distances between every spot of paint increase, in exactly the same way that separations between galaxies increase in the real Universe as it expands. Suppose that the expansion doubles the distance between each spot of paint. Spots that were two centimetres apart end up four centimetres apart; spots that were four centimetres apart end up eight centimetres apart, and so on. If before the expansion there were three spots spaced two centimetres apart in a straight line, then after the expansion the distance from the central spot to either of its neighbours will now be four centimetres, but the distance between the two outer spots will now be eight centimetres. From either of the end spots, the central spot will have receded by two centimetres, but the other end spot will be seen to have receded by four centimetres. It started out twice as far away as the central spot, and the amount of its 'redshift' is twice as big as for the nearer spot. From every spot on the surface of the sphere, the overall picture is the same. Redshift is proportional to distance.

But what if we imagine reducing the size of the sphere? Now, the spots get closer together, and 'blueshift' is proportional to distance. This is equivalent to looking back in time to the history of the expanding Universe. It is obvious that if galaxies are moving apart today then they must have been closer together in the past. It is considerably less obvious, but required by the general theory of relativity, that if you wind this expansion backwards from how things are today for long enough you reach a time when all of the matter and all of space were merged into a mathematical point, a singularity, with zero volume and infinite density, like the singularities predicted to lie at the hearts of black holes. As with black hole singularities, because physicists do not believe theories that predict infinitely extreme physical conditions, it is thought that the general theory must break down when pushed that far.

But there is every reason to believe that the Universe started from a state of extremely small volume (smaller than an atom) and extremely high temperature and density (containing all the mass in the Universe today), even if none of these properties was ever infinite. This idea of a superdense, superhot beginning is the core of the Big Bang model of the Universe. The idea of the Big Bang began to be taken seriously in the second half of the 20th century, as more observations confirmed the reality of the universal expansion. The big question which cosmologists struggled to answer was, when did the Big Bang happen? How old is the Universe? The answer came from studies of galaxies, providing measurements of the Hubble Constant.

The Hubble Constant is a measure of how fast the Universe is expanding today. If it has always been expanding at the same rate, that tells us how long it has been since the Big Bang. Take 1 divided by the value of the Hubble Constant ($1/H$) and you know how long it is since the galaxies were on top of each other – the time since the Big Bang. In the same way, if a car leaves London heading west along the M4 at a steady 60 miles an hour, when it is 120 miles from London we know that the journey started two hours ago. Things are slightly more complicated because the simplest model of the Universe derived from Einstein's equations says that it must have started out expanding more rapidly, and slowed down as time passed, because of gravity holding back the expansion. A better estimate for the age of the Universe is two-thirds of ($1/H$), and $1/H$ itself is referred to now as the Hubble time. But the crucial point is that if we can measure H we can measure the age of the Universe.

Because the age is inversely proportional to H, the smaller the value of the Hubble Constant the older the Universe must be. Using Hubble's own value for the constant, 525 kilometres per second per Megaparsec, the age of the Universe comes out as about two billion years. Even in the 1930s, it was clear that something was wrong with this estimate, because it is less than the age of the Earth. This is one reason why the idea of the Big Bang only began to be taken seriously after the 1940s, when there was a drastic revision of the distance scale after the confusion between different kinds of variables was ironed out. At a stroke, the Hubble Constant was halved and the estimated age of the Universe doubled, making the Universe seem to be about as old as the Earth.

But at about the same time, astronomers began to develop a good understanding of how stars work, and to derive reliable estimates of their ages. Some stars turn out to be more than ten billion years old, which again provided embarrassment for the Big Bang idea as it stood in the 1950s. This was one reason why a rival cosmology, the Steady State model, seemed attractive to some astronomers at the time. The idea behind the Steady State model was that as the galaxies moved apart in the expanding universe, the forces responsible for the stretching of space also caused the appearance of new matter in the gaps between the galaxies – atoms of hydrogen that would form clouds of gas from which new galaxies would form to fill the gaps. On that picture, there had been no beginning and there would be no end, with the universe always having much the same appearance overall. The death knell of the Steady State model was sounded in the 1960s, when radio astronomers discovered a weak hiss of radio noise coming from all directions in space. This cosmic microwave background radiation, which had been predicted by Big Bang theory (although the prediction had been forgotten!), is interpreted as the fading remnant of the energetic radiation from the Big Bang itself, an interpretation reinforced by later observations, including those from dedicated satellites sent in to space to study it. The need for the Steady State alternative also declined because estimates for the age of the Universe gradually increased as the years passed.

From about 1950 onwards, gradual revisions of the distance scale based on ever-improving observations pushed the value of the Hubble Constant down until, by the beginning of the 1990s, it was known to lie somewhere in the range from 50 to 100, in the usual units. As an astronomer would put it, 75 ± 25. This was where the Hubble key project came in.

Like the Andromeda galaxy, galaxies in clusters typically have random motion through space of a few hundred kilometres per hour. This means that in order to get reliable estimates of the cosmological redshift of a cluster it is best to look at distant clusters, where the cosmological redshift is greater and individual random velocities and their associated Doppler shifts are a smaller proportion of the overall redshift. But, of course, it is harder to measure distances for more distant clusters, so there is a trade-off when it comes to using clusters in this way to determine the value of the Hubble Constant. The Hubble key project used the traditional technique devised by Hubble himself of getting accurate distances to nearby galaxies from Cepheids, using these Cepheid distances to calibrate the brightness of other indicators such as supernovae, and moving out into the Universe in a series of steps. The difference was, working 60 years after Hubble, they had a better telescope, the confusion between different kinds of variable stars had been resolved, extinction was understood, and the secondary indicators such as supernovae were also much better understood than in Hubble's day. The final estimate that the key project team came up with for H, in May 2001, was 72 ± 8, corresponding to an age of the Universe of about 14 billion years. Happily, during the previous decade, the 1990s, the ages of the oldest stars we can see were determined by quite independent techniques to be around 13 billion years. The Universe really is older than the stars and galaxies it contains.

This is a much more profound result than it might seem at first sight. The age of the Universe is determined by studying some of the largest things in the Universe, clusters of galaxies, and analysing their behaviour using the general theory of relativity. Our understanding of how stars work, from which we calculate their ages, comes from studying some of the smallest things in the

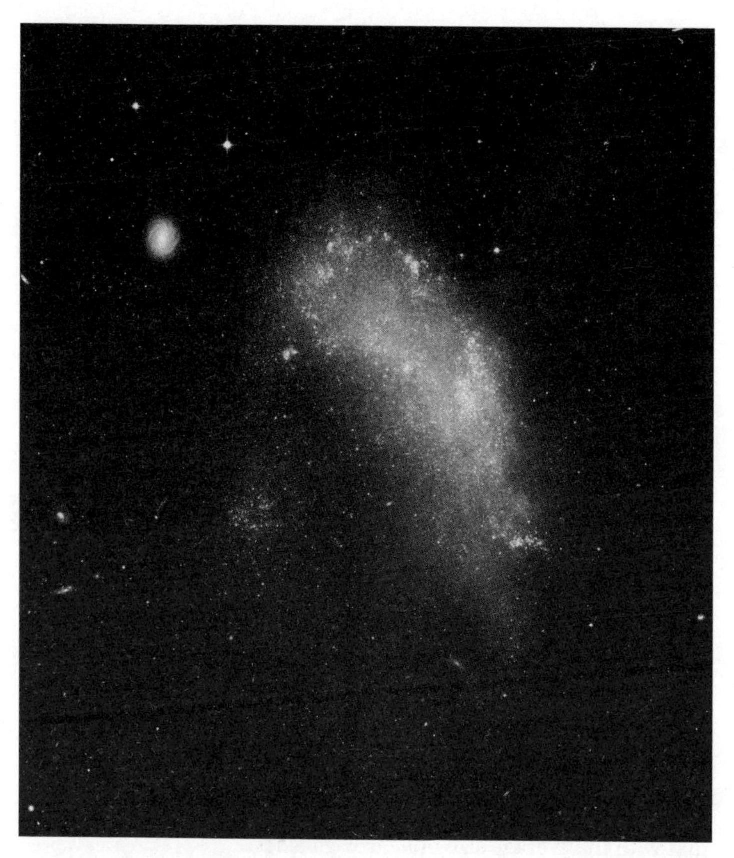

11. The irregular galaxy NGC 1427

Universe, the nuclei of atoms, and using the other great theory of 20th-century physics, quantum mechanics, to calculate how nuclei fuse with one another to release the energy that keeps stars shining. The fact that the two ages agree with one another, and that the ages of the oldest stars are just a little bit less than the age of the Universe, is one of the most compelling reasons to think that the whole of 20th-century physics works and provides a good description of the world around us, from the very small scale to the very large scale.

A value for the Hubble Constant close to 70 kilometres per second per Megaparsec has now been confirmed by other independent techniques. Some of these involve high-tech equipment such as satellites and a sophisticated understanding of physics; but one simple approach highlights the relationship between galaxies and the Universe, and, when combined with the more sophisticated measurements, provides a confirmation of our mediocrity.

The evidence that the Milky Way is just an average spiral is based on a fairly small sample of galaxies fairly close to us, in cosmological terms. If we accept this at face value, though, it provides us with a way to estimate the distances to other galaxies, by comparing their sizes with the size of the Milky Way, or with the average of our local sample, which is near enough the same thing. There is little point in making such comparisons with individual galaxies, because we know that there is a wide range of sizes. The largest spiral galaxy in our cosmic neighbourhood, M101, has a diameter of nearly 62 kpc, more than twice that of the Milky Way, so estimating its distance by assuming it is the same size as our Galaxy would not be a good idea. What we need is some kind of statistical measurement so that we can take the average size of galaxies far away across the Universe and compare that with the average size of nearby galaxies.

Since Hubble's day, observers have built up catalogues giving the positions, redshifts, and angular sizes of thousands of galaxies – many different catalogues each containing thousands of galaxies. Some of these include angular sizes, often given in terms of the same isophotal diameters used to determine the mediocrity of the Milky Way. Each angular diameter can be converted into a true linear diameter by multiplying it by a number which depends only on the redshift, which we know, and H, which we are assuming we do not know. If we take thousands of galaxies with different redshifts, scattered all over the sky, it is possible to choose some value of H and work out all of the linear diameters,

12. The central region of the galaxy M100, imaged by WFPC2 on the HST

then take an average over the whole sample to estimate the average size of a galaxy. It is straightforward to do this over and over again using a computer which keeps on varying *H* until the average value that comes out of the calculation is the same as the average diameter of the nearby spirals like the Milky Way. This gives a unique value for *H*.

There are practical difficulties that have to be overcome. Among other things, you have to make sure that all of the diameters have been measured in the same way, that the sample is restricted to galaxies which have the same overall structure as the galaxies in our local sample, and that the observations are indeed picking up all of the relevant galaxies. One of the most important factors to allow for is that it is easier to see bigger galaxies, so for larger redshifts there will be fewer small galaxies than there should be in the sample because they have been overlooked. This is an effect known as Malmquist bias. Fortunately, by comparing the numbers of galaxies of different sizes at different redshifts it is possible to work out the statistics of this effect – the way small galaxies drop out of the sample as redshift increases – and correct for it. In a further complication, nearby galaxies have to be left out of the calculation, because their random Doppler shifts are comparable to their cosmological redshifts and confuse the picture. But the technique works for galaxies out to about 100 Megaparsecs, and even with all these restrictions one of the standard catalogues, known as RC3, provides a sub-set of well over a thousand suitable galaxies that satisfy all these criteria. This is ample to provide a statistically reliable sample. When all of the work is done, the value for the Hubble Constant based on comparison of galaxy diameters comes out in the high 60s, *if* the Milky Way is indeed just an average spiral. This value agrees with the other measurements.

This is far from being the best or most accurate way to measure the Hubble Constant, but it is valuable for two reasons. First, it is a nice, physical technique which can be understood in terms of our everyday experience, where we know that a cow standing on the other side of a large field only looks small because it is so far away. It does not require any deep understanding of physics or mathematics. Second, the argument can be turned on its head. The first real proof that the Milky Way is just an average spiral came from comparing its size with the sizes of just 17 relatively nearby galaxies. But if *H* is close to 70, as the more sophisticated observations and analyses indicate, then we can use that value to calculate the average size of the 1,000-plus galaxies in our sample, some of them 100 Mpc away, and we find that it is indeed very close to the size of the Milky Way and the average size of our nearby sample. At the very least, our Galaxy is typical of the kind of disc galaxy found in a 'local' region of space some 200 Mpc across with a volume of more than 4 million cubic Megaparsecs.

But this is indeed still a local bubble compared with the size of the observable Universe. There are objects known with measured redshifts corresponding to distances greater than ten billion light years, 30 times farther away than the most distant galaxies used in this technique for estimating the value of H. Studies of these objects show that there is more to the story. It seems that the expansion of the Universe has not slowed down since the Big Bang in the way predicted by the simplest solutions to Einstein's equations, but that it may have begun to speed up.

In the 1990s, astronomers began to use supernova observations to calibrate the redshift–distance relationship for redshifts of about 1 (the largest known redshifts for such supernovae are less than 2). The technique depends on the discovery that a certain kind of supernova, a family known as SN1a, all seem to peak at the same absolute brightness. This was discovered from observations of SN1a in nearby galaxies for which distances are now very well known. The discovery was particularly important because supernovae are so bright that they can be seen at very great distances.

Although all SN1a have the same absolute brightness, the farther away they are across the Universe the fainter they look. This means that if they really do all reach the same absolute peak brightness, by measuring the apparent peak brightness of SN1a in very distant galaxies we can work out how far away those galaxies are. If we can also measure redshifts for the same galaxies, we can calibrate the Hubble Constant. When these observations, at the very limit of what was technologically possible, were carried out, the observers found that the supernovae in very distant galaxies are a little fainter than they should be, if the galaxies in which they

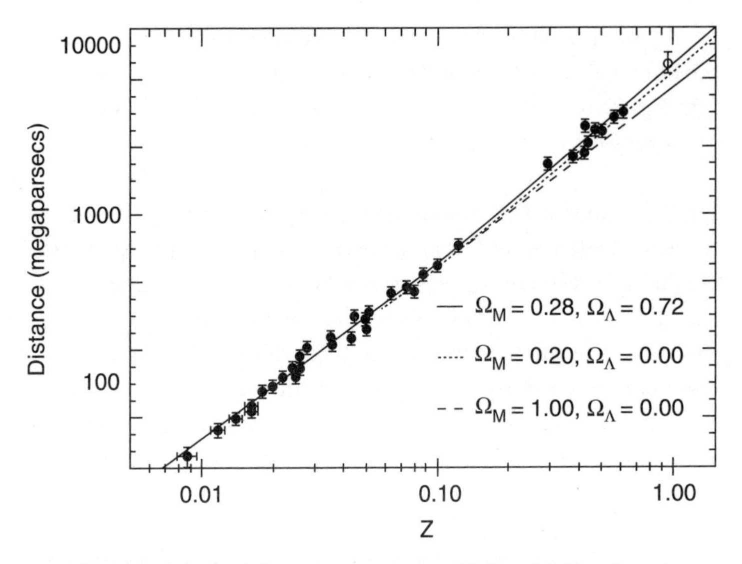

13. Using observation of supernovae at very high redshifts, the redshift–distance plot can be extended far out into the Universe. The best fit to the data (solid line) includes an allowance, *Λ*, for the cosmological constant described previously in the text

reside are at the distances indicated by the accepted value of the Hubble Constant.

The possibility that supernovae in such distant galaxies really do not shine as brightly as those in galaxies closer to us cannot be ruled out, but the conclusion that best fits all of the available evidence is that these supernovae are a little bit farther away from us than they should be if the Universe has been expanding in line with the simplest cosmological models ever since the Big Bang. Just one tiny modification to Einstein's equations is needed to make everything fit – a small cosmological constant has to be put back in to the equations. Perhaps it wasn't such a blunder after all.

When Einstein introduced his cosmological constant, he did so to hold the model universe still. But different choices of the constant will make it expand faster or slower, or even make it collapse. The presence in the equations of the kind of cosmological constant required to explain the supernova observations implies that the entire Universe is filled with a kind of energy which has no noticeable local effect on everyday matter, but acts like a kind of compressed elastic fluid, pushing the Universe outward and acting against the inward pull of gravity. Because the cosmological constant is traditionally labelled Λ, this is known as the lambda field. With a suitable choice of density for this field, it is straightforward to explain how the expansion of the Universe slowed down for the first few billion years after the Big Bang, as the simpler models predicted, but then began very slowly to speed up.

It works like this (there are more complicated possible explanations for the cosmic acceleration, but since the simplest explanation works beautifully I shall not discuss them here). The lambda field is constant, and has had the same value since the Big Bang. Because we cannot see this field, it is often called 'dark energy'. Dark energy is a property of spacetime itself, so when space stretches and there are more cubic centimetres to fill, the dark energy does not get diluted. This means that the amount of energy stored in every cubic centimetre of space stays the same, and it always exerts the same outward push in every cubic centimetre. This is quite different to what happens to matter as the Universe expands. As the Universe emerged from the Big Bang, the density of matter was as great everywhere as the density of an atomic nucleus today. A thimbleful of such material would contain as much mass as all the people on Earth today put together, and so the gravity associated with that density of matter completely overwhelmed the lambda field. As time passed, the Universe expanded and the same amount of matter occupied an increasing volume of space; the density of matter declined accordingly. This meant that the gravitational influence on the expansion gradually got smaller, until it was less than the influence of the dark energy.

To explain the supernova observations, the influence of matter on the expansion, acting to slow the expansion down, must have weakened to the point where it was the same size as the influence of dark energy, acting to make the expansion speed up, about five or six billion years ago. In redshift terms, the switch took place between a redshift of 0.1 and a redshift of 1.7. Since then, the influence of dark energy has been bigger than the influence of matter, making the expansion of the Universe accelerate.

If the expansion is accelerating, one implication is that the Universe is very slightly older than the 14 billion years calculated assuming no acceleration, because if the Universe was expanding more slowly in the past it would have taken longer to reach its present state. But this effect is very small, and it works in the right direction to keep the age of the Universe bigger than the ages of the oldest stars, so it need not concern us here.

The amount of dark energy required to do all this is tiny. Bearing in mind Einstein's discovery that energy and mass are equivalent to one another, the amount of mass associated with dark energy is a bit less than 10^{-29} grams in every cubic centimetre of the Universe – that is, 0.00000000000000000000000000001 grams in every cubic centimetre. So it cannot make the Earth, or the Solar System, or the Milky Way, or even a cluster of galaxies expand and break apart, because on a local scale the gravity of concentrations of matter completely overwhelms it.

On a cosmic scale, though, the presence of even this much energy, and its equivalent mass, in *every* cubic centimetre of the Universe, even in all the 'empty space' between the stars and galaxies, adds up dramatically. It means that there is far more mass in the form of dark energy than there is in the form of bright stars and galaxies. This would have come as a big surprise to Hubble and his contemporaries, who imagined that they were studying the most important components of the Universe. But at the end of the

1990s it was just what the doctor ordered. By then, it had already become clear that there is more to the Universe than meets the eye, and cosmologists were already trying to find what they called the 'missing mass'. The lambda field turned out to be the missing piece that completed the modern picture of the Universe, which provides a framework within which to understand the origin and evolution of galaxies – which are, after all, still very important for life forms like ourselves.

Chapter 6
The material world

What are galaxies made of? The obvious constituents are hot, bright stars and cool, dark clouds of gas and dust. This is essentially the same kind of material that the Earth is made of, and our own bodies are made of – atomic material. Atoms consist of dense nuclei, composed of protons and neutrons, surrounded by clouds of electrons, with one electron in the cloud for each proton in the nucleus. Inside stars, the electrons are stripped from the nuclei to make a form of matter known as plasma, but it is still essentially the same sort of stuff. Protons and neutrons are members of a family of particles collectively known as baryons, and the term 'baryonic matter' is often used by astronomers to refer to the stuff that stars, gas clouds, planets and people are made of. Electrons are members of a different family, known as leptons. But since the mass of an electron is less than one-thousandth of the mass of either a proton or a neutron, in terms of mass, baryons dominate this kind of familiar matter.

One of the remarkable achievements of modern cosmology is that it is able to tell us how much baryonic matter there is in the Universe – or rather, what the density of such matter, averaged over the entire visible Universe, must be. Drawing on the general theory of relativity, cosmologists measure such densities in terms of a parameter labelled with the Greek letter omega (Ω), which is related to the overall curvature of space. This is most easily understood by making an analogy between the three-dimensional curvature of space and the way a two-dimensional surface can be curved. The surface of the Earth is an example of a closed surface, which is bent around on itself. On such a closed surface, if you travel in the same direction for long enough you get back to where you started. The shape of a saddle is an example of an open surface, which can be extended off to infinity in all directions. Exactly in between these two possibilities there is a flat surface, like the top of my desk, which is not curved at all. Einstein's equations tell us that, depending on how much matter it contains, the shape of our three-dimensional space may either be closed, in the same sense that the two-dimensional surface of a sphere is closed, open, like a saddle surface, or flat, like the top of my desk. A flat universe corresponds to having a value of 1 for the density parameter Ω. A closed universe requires a higher density of matter, an open universe a lower density of matter. Cosmologists measure densities as fractions of this parameter. For example, if the amount of baryonic matter in the Universe were half the

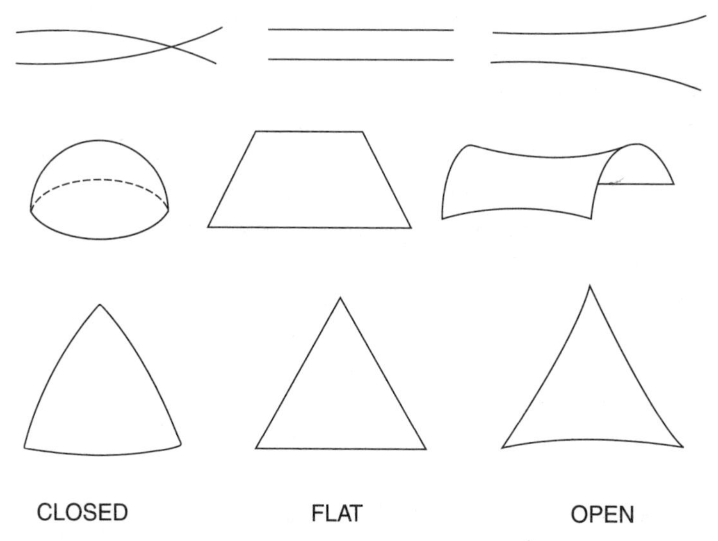

14. Space may conform to one of three basic geometries. These are represented here by their equivalents in two dimensions

amount required to make the Universe flat (which it is *not*) then we would say Ω(baryon) = 0.5.

All the baryonic matter in the Universe was manufactured in the Big Bang, ultimately out of pure energy in line with $E = mc^2$, which can, of course, be rewritten as $m = E/c^2$. The calculation of the amount of baryonic matter produced in the Big Bang is very straightforward, provided that we can be sure that the temperature of the Big Bang was at least a billion degrees. The evidence for this comes from the weak hiss of radio noise that can be detected coming from all directions in space. This background radio noise is interpreted as the leftover radiation from the Big Bang fireball itself, redshifted by a factor of a thousand so that it now shows up as microwave radiation with a temperature of 2.7 degrees above absolute zero (2.7 K). From the observations, we can work backwards to calculate the temperature of the Universe at any time in the past, when it was smaller and the radiation was correspondingly less redshifted. One second after the beginning of time, the temperature was 10 billion K, 100 seconds after the beginning it was 1 billion K, and after an hour it had cooled to 170 million K. For comparison, the temperature at the heart of the Sun is about 15 million K.

Under such conditions, matter is in the form of a plasma, like the inside of the Sun, and radiation gets bounced around between the electrically charged particles. The cosmic microwave background radiation itself comes to us from a time about 300,000 years after the beginning, when the Universe cooled to a few thousand K, roughly the temperature of the surface of the Sun today. Then, negatively charged electrons and positively charged protons got locked up in neutral atoms and the radiation could stream away through space, just as it streams away from the surface of the Sun.

Conditions in the later stages of this cosmic fireball were very similar to the conditions inside exploding nuclear bombs, which have been studied on Earth. Armed with an understanding of how nuclear explosions work, cosmologists can calculate that the baryonic mix that emerged from the Big Bang was about 75 per cent hydrogen and 25 per cent helium by weight, with just a tiny trace of lithium. But from the way the baryonic particles interact with light under extreme conditions, and measurements of the background radiation, they can also calculate that the total amount of baryonic material produced in the Big Bang and present in the Universe is only 4 per cent of the density required for flatness. In other words, $\Omega(\text{baryon}) = 0.04$.

The obvious next step is to compare this prediction of the amount of baryonic matter in the Universe with the amount we can see in bright stars and galaxies. This is a rough and ready calculation based on our understanding of the brightnesses and masses of stars and the number of stars in galaxies, but it suggests that about a fifth of the baryonic matter, less than 1 per cent of the total amount of matter needed to make the Universe flat, is in the bright stuff. Some of the other four-fifths is in the clouds of gas and dust between the stars, or perhaps in the form of dead, burnt-out stars. Some of it is in the form of a kind of transparent fog of hydrogen and helium surrounding galaxies like our own. And yet, as I mentioned earlier, we know from the way galaxies rotate and the way they move through space that they are held in the grip of a great deal more matter than this. This can only be a form of cold, dark, non-baryonic matter, made up of some kind of particle or particles that have never been detected in any experiment on Earth. It is dubbed Cold Dark Matter, or CDM for short, and detecting it is one of the most pressing tasks of particle physicists today.

Evidence for CDM comes from the way galaxies move – how they rotate, and how they move through space. The rotation of a disc galaxy can be measured using the familiar Doppler effect, which shows how stars on one side of a galaxy are moving towards us as the galaxy rotates, while stars on the other side are moving away from us. This only works for galaxies seen nearly edge-on,

15. The microwave map of the sky obtained by WMAP

but there are plenty of those to study. The Doppler effect adds to the redshift on one side of the disc, and subtracts from it on the other side, so the measured redshift at different places along the disc shows how the stars are moving around the centre of the galaxy. The crucial point is that outside the central nucleus of a disc galaxy, where other interesting things happen, the rotation speed is constant all the way out to the edge of the visible disc. All the stars in the disc are moving at the same speed in terms of kilometres per second. This is quite different from the way the planets of the Solar System move in their orbits around the Sun.

Planets are small objects orbiting a large central mass, and the gravity of the Sun dominates their motion. Because of this, the speed with which a planet moves, in kilometres per second, is inversely proportional to the square of its distance from the centre of the Solar System. Jupiter is farther from the Sun than we are, so it moves more slowly in its orbit than the Earth, as well as having a larger orbit. But all the stars in the disc of a galaxy move at the same speed. Stars farther out from the centre still have bigger orbits, so they still take longer to complete one circuit of the galaxy. But they are all travelling at essentially the same orbital speed through space.

This is exactly the pattern of behaviour that corresponds to the orbital motion of relatively light objects embedded within a large amount of gravitating matter, like raisins moving around inside a loaf of raisin bread. The natural conclusion is that disc galaxies, including the Milky Way, are rotating inside much larger clouds, or haloes, of unseen dark matter. This is some form of spread-out material, so it must be in the form of particles rather like the particles of a gas, which have mass and influence everyday matter gravitationally, but do not interact with everyday matter in any other way (for example, through electromagnetism) or they would have been noticed. On this picture, CDM particles are present everywhere, including the place where you are reading this, and

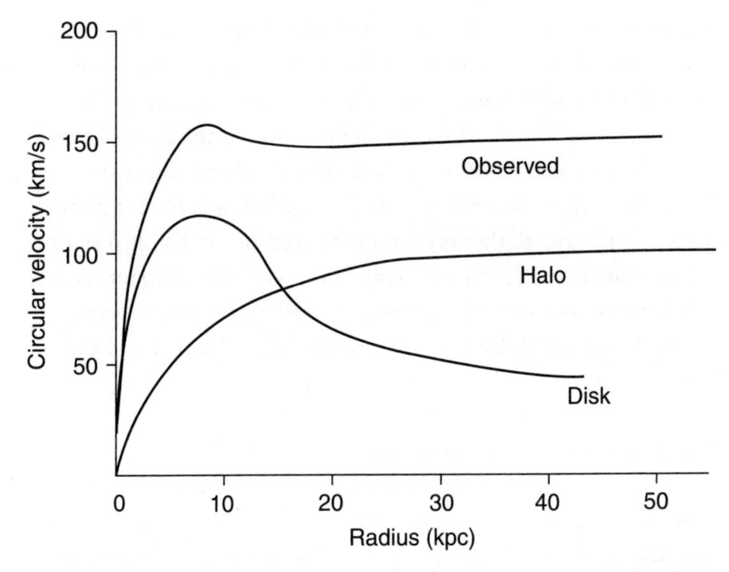

16. A schematic representation of the typical 'rotation curve' seen in a disc galaxy

are continually passing through your body without affecting it. There are thousands, perhaps tens of thousands, of CDM particles in every cubic metre of everything, as well as in every cubic metre of 'nothing' – so-called empty space.

Cold Dark Matter also reveals its presence through its influence on clusters of galaxies. The invaluable Doppler shift can be used to tell us the way an individual galaxy in a cluster is moving relative to the centre of the cluster, and the range of speeds of all the galaxies in a cluster. The clusters can only exist because they are held together by gravity – otherwise, the expansion of the Universe would pull them apart and spread the galaxies out through space. But there are limits to how effective this gravitational constraint can be. If you throw a ball up into the air, it will reach a certain height, depending on its speed, then fall back as the Earth's gravity tugs on it. But if you could throw the ball hard enough it would escape from the Earth entirely and carry on out into space. The minimum vertical speed needed to do this is called the escape velocity, and depends only on the mass of the object you are trying to escape from and how far you are from the centre of the mass. At the surface of the Earth, the escape velocity is 11.2 km per second. If we add up the masses of all the galaxies in a cluster, inferred from their brightnesses *and* including an appropriate allowance for their dark matter haloes, we can work out the escape velocity from the cluster. It turns out that in order for clusters to maintain their gravitational grip on their galaxies there must be even more dark matter in the 'empty space' between the galaxies, as well as the dark matter in the haloes of individual galaxies. The whole Universe is filled with an invisible fog of CDM.

Putting all of the evidence together, it can be calculated that there is nearly six times as much Cold Dark Matter in the Universe as there is baryonic matter. In other words, $\Omega(\text{CDM}) = 0.23$. Adding this to the known amount of baryonic matter in the Universe, we find that 27 per cent of the amount of matter needed to make the Universe flat has been accounted for. That is, $\Omega(\text{matter}) = 0.27$.

This could have been embarrassing for cosmologists, because by the time these calculations were being refined to the accuracy I have given here, around the end of the 20th century, there was other evidence that the Universe is actually flat. It came from studies of the cosmic microwave background radiation, made by instruments carried on balloons and satellites above the obscuring layers of the Earth's atmosphere. Such instruments are now so sensitive that they can pick out variations in the temperature of the radiation from place to place on the sky, looking at hot and cold spots (relatively speaking) that were imprinted on it when the Universe was a few hundred thousand years old.

Before the Universe cooled to the point where electrically neutral atoms could form, radiation and the electrically charged particles of matter were coupled with one another in such a way that differences in the density of matter at different places in the Universe were associated with differences in the temperature of the radiation. About 300,000 years after the Big Bang, when the Universe cooled to the critical temperature, radiation and matter decoupled, and the radiation was left imprinted with a pattern of hot and cold spots corresponding to the pattern of density variations in baryonic matter at that time – a kind of fossil of the large-scale distribution of baryons at the time of decoupling. Because light travels at a finite speed, in 300,000 years it can only travel a distance of 300,000 light years, so in the time from the Big Bang until decoupling the largest regions of the Universe that could have had any kind of internal coherence grew to be 300,000 light years across. This means that the biggest uniform patches seen in the microwave background map of the sky correspond to patches of the Universe that were 300,000 light years across at the time of decoupling.

Since that time, the radiation has streamed across space without interacting directly with matter. But it has been influenced by the curvature of space. We know that a massive object like the Sun bends light passing near its edge. This is very similar to the way a lens bends light. Lenses can make images of distant objects seem bigger (like looking through a telescope) or smaller (like looking through the wrong end of a telescope). So can curved spacetime, depending on the nature of the curvature. Using the general theory of relativity, it is possible to calculate how big the largest uniform blobs in the background radiation should look to our instruments today, if they were 300,000 light years across at the time of decoupling. The observed size depends on the exact curvature, but if the Universe is open we should see a magnification and if it is closed we should see smaller blobs. If it is flat, there should be no effect. The measurements show that the Universe is almost certainly flat, but might *just* be closed. In other words, $\Omega = 1$.

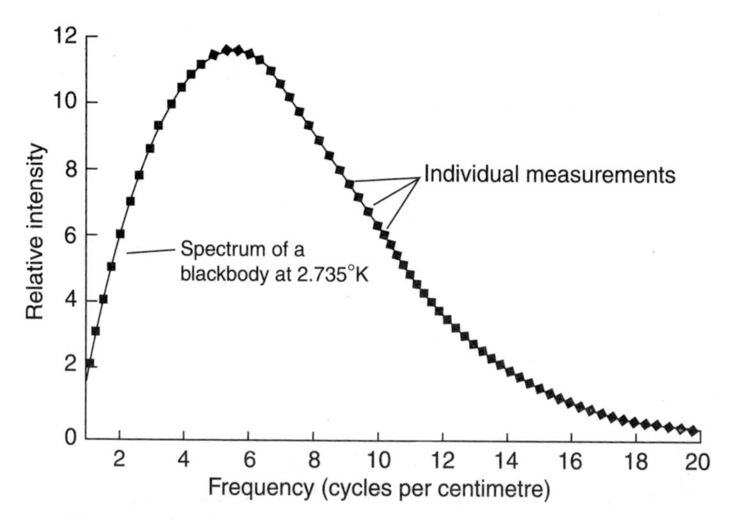

17. The spectrum of cosmic background radiation measured by the COBE satellite

And yet, we know that the total amount of matter in the Universe is less than one-third of the amount needed to make the Universe flat. It could indeed have been embarrassing. But just at the time cosmologists were beginning to worry about this puzzle, the supernova studies came along and showed that the expansion of the Universe is accelerating. The amount by which it is accelerating requires a cosmological constant Λ with a certain strength. This corresponds to a mass density equivalent of 73 per cent of the density of matter needed to make the Universe flat. In other words, $\Omega(\Lambda) = 0.73$. This was just what was needed. Far from being an embarrassment, the discovery that $\Omega(\text{matter}) = 0.27$ turns out to be a triumph. When everything is taken into account, we are left with an equation that is very simple, and very true:

$$\begin{aligned}\Omega &= \Omega(\text{baryon}) + \Omega(\text{CDM}) + \Omega(\Lambda)\\ &= 0.04 + 0.23 + 0.73\\ &= 1\end{aligned}$$

As Mr Micawber would have said, 'result, happiness'. For obvious reasons, this package is known as 'ΛCDM' cosmology, and it is one of the great triumphs of science.

The next phase in developing our understanding of the Universe, a work still in progress, is to account for the origin of the kind of galaxies we see in the Universe within the framework of ΛCDM cosmology. But before we can do this, we need to take stock of the material world – the different kinds of galaxies we have to explain – since, unfortunately, this does not just consist of a neat division into disc galaxies and ellipticals.

The visible parts of spiral galaxies like the Milky Way form the classic two-part structure of a disc and a central nuclear bulge, although in some cases the bulge is very small. The spiral arms are the most visible characteristic of the disc, but the large quantity of dust and gas is just as important because it provides the raw material for the formation of the hot, young stars of the disc, known as Population I. The stars of the bulge and the globular clusters around a disc galaxy are older Population II stars. Spirals come with and without central bars, which may be temporary features that all spirals grow at some time. Most bright galaxies are spirals, and it is now accepted that all disc galaxies have black holes at their hearts, like the one at the centre of the Milky Way. The largest spirals may contain as many as 500 billion stars.

Disc galaxies without spiral arms (sometimes known, for historical reasons, as lenticular galaxies) still have the basic disc and bulge structure, but lack the dusty clouds. They are mostly made of Population II stars, and the inference is that they have used up all their star-forming material and settled into a quiet middle age. Distant lenticular galaxies seen at various angles can hardly be distinguished from ellipticals, although if their rotation can be measured by the Doppler effect that is a sure indication of their true nature.

Elliptical galaxies do not rotate as a whole, but the individual stars in ellipticals orbit around the centre of the galaxy. In nearby ellipticals that can be studied in detail it is possible to pick out streams of stars following different orbits oriented in many different directions, like the star streams in the Milky Way but on a grander scale. This variety of differently oriented star streams is what gives elliptical galaxies their overall shape, which is, strictly speaking, ellipsoidal, like a squashed or stretched sphere. They are dominated by old, Population II stars, and superficially look rather like the bulge of a disc galaxy without the disc. At least some ellipticals do contain dust, often in rings around the centre of the galaxy, but star formation is not going on in them in a major way at present. Although most bright galaxies are spirals, the largest galaxies are giant ellipticals which contain more than a trillion stars and are hundreds of kiloparsecs across. But the smallest galaxies in the Universe also seem to be ellipticals, containing only a few million stars and typically only a kiloparsec or so across. The smallest of these dwarf galaxies are comparable in size to the largest globular clusters, which is probably a clue to the origin of globular clusters. We can only see such tiny galaxies in our neighbourhood, where half of the couple of dozen nearest galaxies are dwarf ellipticals. It is very likely that most of the galaxies in the Universe are dwarfs like this, but we cannot see them at great distances.

Anything which cannot be described as an elliptical or as a disc galaxy is classed as an irregular galaxy. Irregulars usually contain a lot of gas and dust, in which very active star formation is going on. Because there is no well-defined structure like that of a spiral, this produces patches of star formation dotted around the galaxy, giving it an irregular, patchy appearance on photographs. The Magellanic Clouds, two small galaxies in the gravitational grip of the Milky Way, used to be classified as irregulars but have now been found to have an underlying barred spiral structure, difficult to see because of the patchy nature of the star formation. Some irregulars may be remnants or pieces of larger galaxies that have been disrupted tidally by close encounters with other galaxies. Such close encounters can be seen occurring across the Universe. In some cases, galaxies can be seen passing close by one another, being stretched and distorted by tidal forces; in other examples, galaxies are colliding with one another and may be in the process of merging – an important clue, as we shall see, to the origin of the kinds of galaxies we see around us.

Encounters between galaxies can also trigger massive bursts of star formation, which astronomers refer to, prosaically, as starbursts. There is no formal definition of a starburst galaxy, but it is one in which the rate at which stars are being formed is so great that all of the available gas and dust would have to be used up in a time much shorter than the age of the Universe. So they must be transient phenomena. In some starburst galaxies, stars are forming at a rate of hundreds of solar masses a year, about a hundred times faster than the rate of star formation in our own Galaxy. This would typically use up all the available material within about a hundred million years, less than 1 per cent of the age of the Universe.

Some starburst galaxies, especially the smaller ones, appear very blue, because the light from them is dominated by hot, young blue stars. These galaxies contain little dust, presumably the consequence of having recently been disturbed by an interaction or merger with another system, which stirred the dusty gas clouds up and triggered the burst of star formation which has depleted these reservoirs. Individual bursts of star formation occur within these galaxies in compact clusters of stars up to 20 light years (6 or 7 parsecs) across, a hundred million times brighter than our Sun. At the other end of the scale, some starburst galaxies are very large and very red, and are detected at infrared wavelengths using instruments carried into space on satellites. This is because they are enveloped in huge quantities of dust, which absorbs the light from the young stars inside the galaxy and reradiates it at infrared wavelengths. X-ray telescopes see right through the dust, and reveal that many of these large starburst galaxies have double cores of activity. This suggests that they are the result of two large galaxies merging. The double core is formed from the two black holes, one from each of the merging galaxies, which have not yet themselves merged. Starburst galaxies were found to be common, once astronomers had the technology to look for them and knew what they were looking for.

The presence of black holes at their hearts also explains why some galaxies show signs of violent activity in their nuclei, with outbursts flinging material away into space. Such objects were discovered piecemeal over the course of many decades, using different kinds of telescopes observing in different parts of the electromagnetic spectrum – visible light, radio, infrared, X-rays, and so on. As a result, many different names were given to these objects, which are now thought to be members of a single family. The generic name 'active galactic nucleus', usually shortened to AGN, therefore embraces a variety of such objects going under names such as Seyfert galaxies, N galaxies, BL Lac objects, radio galaxies, and quasars. It is now thought that these are all powered by the same kind of process, involving matter falling in (or on) to a supermassive black hole, with the differences being only those of degree, not of kind.

When material falls onto a black hole, the gravitational energy associated with it is released, being turned into energy of motion (kinetic energy) as the material speeds up. The same thing happens on a smaller scale if you drop something out of an upstairs window. The object falls downward at an increasing velocity as gravitational energy is converted into motion; then, when it hits the ground the kinetic energy is converted into heat, shared out among the molecules in the ground, which move a little faster as that patch of ground warms slightly. The 'Hot Spot' technology used in TV broadcasts of sporting events such as cricket matches makes use of this technology to show exactly where a ball has struck.

The particles of matter in the stuff falling into a black hole also collide with one another and get hot as they try to funnel into the hole, forming a swirling disc of hot material known as an accretion disc. The gravitational field of a black hole is so intense that a great deal of energy can be released in this way – up to 10 per cent of the mass energy, mc^2, of the infalling material. If the core black hole in a galaxy has a mass of only a hundred million times the mass of our Sun, roughly 0.1 per cent of the mass of all the bright stars in the surrounding galaxy put together, then it would only need to swallow the equivalent of a couple of stars like the Sun each year to provide the energy output seen in the most active AGN.

All large galaxies probably go through a phase of such activity, settling down into quiet respectability, like the Milky Way, when all of the 'fuel' near to the central black hole has been swallowed. But they could be reactivated if an encounter with another galaxy shakes things up enough for a fresh supply of gas and dust, or even stars, to spiral into the black hole. Any stars that suffer this fate get ripped apart by tidal forces into their component particles long before they are swallowed.

The energy from the central source is often beamed out in two directions on opposite sides of the galaxy. This is probably because the accretion disc of material around the black hole prevents energy escaping along the 'equator'. Both matter and energy can be ejected from the central region of the galaxy as a result, sometimes forming thin jets which interact with their surroundings to produce lobes of radio noise on either side of the galaxy. The most active AGN, the class known as quasars, are so bright that it is very difficult – sometimes impossible – to see the stars of the surrounding galaxy in their glare. As a result, they look like stars in ordinary photographs and their true nature is only revealed by measuring their redshifts. They radiate typically as much as 10,000 times as much energy as all the stars in the Milky Way put together, and some can be seen, even using optical telescopes on the surface of the Earth, at distances greater than 13 billion light years, with redshifts greater than 6; many are known with redshifts bigger than 4, corresponding to a distance of some 10 billion light years. But quasars are exceptionally bright, and are not necessarily typical of their surroundings; happily, large numbers of much fainter distant objects, relatively quiet galaxies even closer in time to the Big Bang, have been detected using the Hubble Space Telescope (HST) pushed to its limits.

The importance of studying objects at great distances across the Universe is that when we look at an object that is, say, 10 billion light years away, we see it by light which left it 10 billion years ago. This is the 'look back time', and it means that telescopes are in a sense time machines, showing us what the Universe was like when it was younger. The light from a distant galaxy is old, in the sense that it has been a long time on its journey; but the galaxy we see using that light is a young galaxy. Early studies of quasars showed that they were more common when the Universe was younger, just as you would expect if they are powered by accretion and fade away when they have swallowed all the available material. Historically, this was one of the clues that tilted the balance of evidence in favour of the Big Bang model and away from the Steady State idea. But the deepest observations made by the HST, corresponding to a look back time in excess of 13 billion years, tell us much more.

There is one further curiosity about all this that should be mentioned. For distant objects, because light has taken a long time on its journey to us, the Universe has expanded significantly while the light was on its way. So although a look back time of, say, 4.25 years implies that we are looking at an object that is 4.25 light years away, a look back time of 4.25 billion years implies that we are looking at an object that was in a sense 4.25 billion light years away when the light started its journey, but is now significantly farther away – in this case, getting on for twice as far. (It is even more complicated than this, since the distance light has to travel starts increasing as soon as it sets out on its journey, but this oversimplification will suffice to make the point.) This raises problems defining exactly what you mean by the 'present distance' to a remote galaxy, not least since nothing can travel faster than light so we have no way of measuring the 'present distance'. So like other astronomers I shall use look back time as the key indicator of how far away an object is, without trying to convert this into distances for anything outside our local region of the Universe. The 'distances' referred to earlier in this chapter should really be regarded as the equivalent of look back times.

Among the many advantages that photographic and electronic recording methods have over the human eye, the most fundamental is that the longer they look, the more they see. Human eyes essentially give us a real-time view of our surroundings, and allow us to see things – such as stars – that are brighter than a certain limit. If an object is too faint to see, once your eyes have adapted to the dark no amount of staring in its direction will make it visible. But the detectors attached to modern telescopes keep on adding up the light from faint sources as long as they are pointing at them. A longer exposure will reveal fainter objects than a short exposure does, as the photons (particles of light) from the source fall on the detector one by one and the total gradually grows. In the most extreme example so far of the application of this process, between 24 September 2003 and 16 January 2004 astronomers exposed the HST for a total of a million seconds to a tiny patch of sky in the constellation Fornax that looks completely black in ordinary photographs. The electronic image-gathering took place in 800 separate exposures, which were stored electronically then combined in a computer to give the equivalent of a single exposure more than eleven days long. The resulting image showed that this seemingly blank piece

18. The Hubble Ultra Deep Field

of sky is filled with galaxies, some of which are seen by light which left them when the Universe was less than 800 million years old, at a redshift of about 7.

The image is known as the Hubble Ultra Deep Field, or HUDF. The patch of sky on the image corresponds to just one thirteen-millionth of the area of the whole sky, no larger than a grain of sand held at arm's length, and has been described by the astronomers involved as equivalent to looking through a drinking straw 2.5 metres long. Yet this tiny patch of sky contains roughly 10,000 galaxies visible in the HUDF image. The ones which are of particular interest here are the faintest and reddest of these galaxies, with the largest look back time. The light from these particular objects trickled in to the detector on the HST at a rate of just one photon per minute.

Although the HUDF contains many normal galaxies, including spirals and ellipticals, these more distant objects have a variety of strange shapes, and some of them are clearly involved in interactions with one another. Some of the galaxies seem to be arranged like links on a bracelet, others are long and thin like toothpicks, and there is a variety of other peculiar shapes. At these early times in the history of the Universe, there were no spirals and no ellipticals – nothing resembling the kind of galaxies in our neighbourhood. Astronomers interpret this as evidence that they have captured a snapshot of the early stages of galaxy formation, before the galaxies settled down into the kinds of regular structures we see in the Universe at more recent times. When they are able to look back even farther in time with the next generation of telescopes, they expect to see nothing at all – the so-called 'dark age' between the time when radiation and matter decoupled, a few hundred thousand years after the Big Bang, and the time when the first galaxies formed, a few hundred million years after the Big Bang. Detecting nothing would, in this case, be a triumphant confirmation of a scientific theory. The oldest objects seen in the HUDF may themselves be at the edge of the dark age, about 400 million years after the Big Bang, at a redshift of about 12.

The most remarkable thing about these galaxies – perhaps we should call them proto-galaxies – is that they exist at all at such times. In much less than a billion years, the Universe had gone from being a sea of hot gas to a place where clumps of matter big enough to grow into the galaxies we see around us already existed, and were holding back, by gravity, matter that would otherwise have been spread ever thinner as the Universe expanded. This can only have happened if there were some kind of seeds on which galaxies could grow, cores with a strong enough gravitational influence to overcome the universal thinning. The identification of those cores with supermassive black holes proved the last link in a model of galaxy formation which explains how galaxies like the Milky Way came to be the way they are, and ultimately, since we are part of the Milky Way Galaxy, why we are here at all.

Chapter 7
The origin of galaxies

Before looking in detail at the explanation of how galaxies got to be the way they are, it makes sense to take stock of the way the Universe looks today, so that we have a clear idea of what it is we are trying to explain. I have already described the nature and appearance of individual galaxies, and mentioned the fact that most galaxies occur in clusters that are held together by gravity. But there is another layer of structure to the Universe, which provides important clues to the origin of galaxies. On the very largest scales, galaxies (strictly speaking, groups of galaxies and small clusters) line up in filaments that criss-cross the Universe and meet each other at intersections where there are large clusters of galaxies. Between the filaments there are darker regions where galaxies are rare. An analogy that is often made is with the view from space of a large, developed part of the world, such as Europe or North America, at night. Roads that cross the country are lit up by street lights and the lights of vehicles, and converge on brightly lit cities; between the roads, the countryside is dark. The key difference is that the distribution of galaxies in the Universe is three-dimensional, forming a foamy-looking structure as seen from Earth, revealed in the latest redshift surveys of the nearby Universe, out to a redshift of about 0.5. Unlike clusters and superclusters of galaxies, these filaments are not gravitationally bound together; extending the analogy with roads, they are simply the routes along which galaxies are moving as different clumps of matter tug on one another. But their existence does reveal how much matter is doing the tugging.

The overall pattern in the distribution of galaxies in three dimensions has now been studied in great detail by teams of astronomers who map the distribution of millions of galaxies on the sky, using redshifts to establish their distances. These observations of the relatively nearby Universe can be compared with the pattern of hot and cold spots seen in the microwave background radiation, imprinted at a redshift of 1,000, and also with computer simulations of how galaxies could grow in a variety of different model universes. The theoretical understanding of the way the Universe began expanding says that during the fireball stage, when baryonic matter and radiation were closely linked together, space was criss-crossed by sound waves with all wavelengths up to the limiting size, mentioned before, set by the speed of light. After decoupling, as we have seen, the radiation still carried an imprint of the pattern made by the sound waves, while the baryons settled down into clumps of matter held together by gravity. By applying statistical techniques to analyse the pattern of galaxies seen in the Universe around us, astronomers have now been able to detect the signature of these sound waves (so-called 'acoustic peaks') in the distribution of matter itself.

In 2005, two teams using different analyses both reported that statistical variations in the distribution of galaxies seen in large three-dimensional surveys show the imprint of these sound waves from the Big Bang. Observationally, everything fits together. But the computer simulations tell us that it would have been impossible for structures as large as the ones we see in the Universe today to have grown from the ripples present in the Big Bang fireball in the time available since the Big Bang, if the only thing pulling the baryons into clumps was their own gravity. The point is that, although the sound waves may have been large in the sense of having a long wavelength, they were also very shallow, merely ripples in the cosmic sea.

19. The simulation of the distribution of matter in the expanding Universe described in the text. This closely matches the observed distribution of galaxies

The need for some extra gravitational influence should hardly come as a surprise, since I have already discussed the evidence for the presence of dark matter from the way individual galaxies rotate and the fact that clusters of galaxies are bound together gravitationally. But this is a completely different piece of evidence for the existence of dark matter, and the computer simulations are so sophisticated that they can tell us precisely how much Cold Dark Matter is needed to do the trick.

Such simulations track the behaviour of individual 'particles' moving under the influence of gravity in the expanding model universe. Each particle corresponds to about a billion times the mass of the Sun, and in the largest simulations to date ten billion particles are involved, moving in accordance with the known laws of physics. The simulations begin with the particles arranged statistically in the same way that we know matter was distributed at the time of decoupling, and then move forward in a series of steps taking account of the way the universe is expanding. The simulations can be chosen to include the effects of different kinds of cosmological constant, different amounts of dark matter, and different values for the curvature of spacetime. The process takes a lot of computer time. To obtain the simulation shown in Figure 19, a cluster of Unix computers using 812 processors with two terabytes of memory performing 4.2 trillion calculations per second ran for several weeks. Overall, the simulation produced a series of 64 snapshots of the model universe at different stages of its development, corresponding to different times since the Big Bang and culminating in the present day.

The results are clear; statistically, the simulation looks just like the real Universe, which is why I chose it. It represents the only class of such models that look like this. Starting from the kind of pattern of irregularities seen in the microwave background radiation, the kind of distribution of galaxies we see in the Universe today can only be produced in 13 billion years if the Universe is flat, there is six times more Cold Dark Matter than there is baryonic matter, and the cosmological constant contributes some 73 per cent of the mass density of the Universe. It is, of course, the highly successful ΛCDM model. The key to the formation of the observed structure is that as soon as baryonic matter decoupled from the radiation and was free to go its own way, in regions of the early Universe where there was already a slightly greater density of dark matter this pulled the nearby baryonic gas into the gravitational equivalent of potholes, where clouds of gas became dense enough to collapse and form galaxies and stars, distributed in a foamy pattern across the Universe. In the dark voids between the bright filaments there is still nearly the same density of both baryons and CDM, and it only needed a small (that is, shallow) ripple here and there to create the conditions needed to get the gas clouds to collapse. Changing the analogy from the road network mentioned earlier, the bright filaments can be thought of as rivers along which baryons flow.

This is the framework within which astronomers now believe they have a good understanding of the way in which individual galaxies have formed.

Just after decoupling, the baryonic material was still far too hot to collapse very much even in the presence of dark matter. But, crucially, the dark matter, being cold, began to collapse immediately in places where the density was a little higher than the average. Until about 20 million years after the Big Bang, corresponding to a redshift of about 100, the Universe was still pretty smooth, but the Cold Dark Matter particles were beginning to pull themselves into gravitationally bound clumps capable of holding matter back against the outward expansion of the Universe. Starting from the kind of ripples present in the background radiation, by a redshift of about 25 to 50 the dark matter would have formed clumps containing about the same mass as the Earth but about as big across as our Solar System. Most of the mass of such spherical clouds was concentrated near the centre, and the clouds formed in this way had a strong enough gravitational influence on each other to resist the universal expansion and form clusters, clusters of clusters, and so on in a hierarchical 'bottom–up' structure. This brought baryonic material streaming down onto the greatest concentrations of mass, forming stars and then galaxies at the nodes of the filaments as it did so and producing the filamentary 'cosmic highway' appearance of the Universe.

The first bright objects to appear in the Universe would have been massive stars, with between a few tens and a few hundred times as much mass as our Sun. These would have been very different from the stars around us today, because they contained only hydrogen and helium produced in the Big Bang, with none of the heavier elements. The first star-forming systems would have been part of a local filamentary structure which gradually became a subcomponent of a bigger filamentary structure stretching in a hierarchical fashion across the Universe, and still developing as clusters and superclusters of galaxies stream together in filaments. The models suggest that star-forming regions appeared about 200 million years after the Big Bang, each containing between a hundred thousand and a million times as much mass as the Sun, and between 30 and 100 light years across, similar in size to the clouds of gas and dust in which stars are forming today in the Milky Way. But these 'clouds' consisted chiefly of dark matter.

Simulations of the way baryons could coalesce to form stars in such clouds suggest that a filamentary structure like the large-scale filamentary structure developed inside each cloud, with matter concentrating at the nodes of the filaments. As the density increased, collisions between atoms became more common and some hydrogen atoms would have got together to make molecules of hydrogen. These molecules, crucially, would cool the gas inside the cloud by emitting infrared radiation; molecules of helium would do the same thing, but less efficiently. It was only this cooling that allowed the baryonic gas in the cloud to collapse still further into proto-stars, separating out the baryons to some extent from the dark matter.

In star-forming regions today, the cooling process is much more efficient thanks to the presence of heavier elements, which is why the clouds are able to collapse as much as they do before stars form. But in the primordial star-forming clouds everything happened at a higher temperature, with the result that the first star-forming knots in the cloud still had masses of a few hundred to a thousand solar masses. Just as in the case of stars forming today, it was very difficult for these clouds to fragment, and each cloud could have formed only a few (probably no more than three) stars, with some of the mass being blown away in winds as the proto-stars warmed up.

The result would have been a first population of stars (confusingly dubbed 'Population III' from an extension of the traditional nomenclature for stars in our Galaxy) with masses typically of a few hundred times the mass of the Sun and surface temperatures of about 100,000 K, radiating strongly in the ultraviolet part of the spectrum. This radiation, which filled the early Universe, is still visible today, but as a result of redshifting as an infrared glow detected by the Spitzer Space Telescope.

Although the first stars were bright, they would have been short-lived. The lifetime of a star depends inversely on its mass, because massive stars have to burn their fuel more vigorously to hold themselves up against their own weight. Within a few million years, still only about 200–250 million years after the Big Bang, stars which started out with masses roughly in the range from 100 to 250 times the mass of the Sun would have exploded completely at the end of their lives, spreading their material throughout the surrounding gas clouds. This material included the first heavy elements, which made cooling much more efficient when the next generation of stars formed, enabling the star-forming condensations in regions triggered into collapse by the blast waves from the exploding stars to become much smaller and make the first stars comparable to those in the Milky Way today. Indeed, some of those second generation stars may still be present in our Galaxy – the oldest Population II stars are calculated to have ages in excess of 13.2 billion years, so they formed within about 500 million years of the Big Bang.

Stars which have masses in excess of about 250 times that of the Sun are not completely disrupted in an explosive death. Instead, most of the material they contain collapses to make a black hole. These primordial stars formed in the densest concentrations of matter in the Universe at that time, so it is likely that the black holes would be close enough to each other for mergers to take place and the black holes to grow into even more massive objects. Nobody can be quite sure where the supermassive black holes at the hearts of galaxies today came from, but it seems at least possible that this merging of black holes left over from the first generation of stars began the process by which

20. A black hole at work. The jet emerging from the centre of the galaxy M87 is powered by a black hole. Although just visible in an optical photograph (left), the jet shows up much more clearly in the infrared (right)

supermassive black holes, feeding off the matter surrounding them, formed.

Observations of quasars at redshifts around 6.5 show that black holes with at least a billion times the mass of the Sun had formed long before the Universe was a billion years old. These are exceptionally large examples, which is why those quasars are bright enough to be seen at look back times close to 13 billion light years, but they confirm the speed with which galaxies appeared. Simulations show that there must have been many lesser black holes as well, forming the cores on which galaxies grew; each black hole may have been embedded in a halo containing a thousand billion solar masses of material. Baryonic material fell into the black hole, giving up gravitational energy to power quasars and other AGN, while stars formed in the quieter outer regions of what had become a galaxy as the baryonic material settled down; but the simulations also show that very large numbers of the original Earth-mass dark matter clouds should have survived all of this turmoil right up until the present day, and still be present in the dark matter haloes around the galaxies. It is estimated that there may be a thousand trillion (10^{15}) such objects in the halo of our Galaxy alone.

The calculations show that the process I have described can form an object as big as the Milky Way in the time available – a few billion years – provided that the central black hole has a mass of at least a million solar masses. Happily, observations reveal that the mass of the black hole at the heart of the Milky Way is about three million times the mass of our Sun. Everything fits. But although astronomers have a self-consistent model of how the first galaxies formed, there is still more to be explained, including an intriguing correlation between the mass of the black hole at the heart of a galaxy and the properties of the surrounding galaxy.

It is worth remembering how new the study of supermassive black holes of this kind is. They can only be studied directly in relatively nearby galaxies, where the presence of a massive central object is revealed by measuring the speeds of the stars orbiting near it, using the Doppler effect. The first supermassive black hole was only identified in 1984, and from then until the end of the 20th century simply finding one was an event; there were nowhere near enough of them known to make generalizations about their properties. But by the year 2000, the number of known supermassive black holes was up to 33, and one or two more are found each year. This is enough to begin to attempt an understanding of the relationship between such objects and their host galaxies.

At the beginning of the 21st century, astronomers discovered a relationship between the mass of the central black hole in a galaxy and the mass of the bulge of stars at the centre of the disk, or in the case of ellipticals the mass of the whole galaxy. There is no correlation with the properties of the disk itself; disks seem to have been added as an afterthought following the development of the bulge. Since the bulge at the centre of a disc galaxy closely resembles an elliptical galaxy, it seems likely that all primordial elliptical galaxies initially grew around black holes in the same way, but that not all of them then developed discs, possibly because of a lack of raw material from which a disc could form. So when referring to the generic properties of ellipticals and the bulges of disc galaxies, astronomers use the term 'spheroid'.

The masses of the supermassive black holes are determined by measuring the velocities of stars very close to the centre of a spheroid. The mass of the spheroid can be estimated from its brightness. But it is also possible to calculate the average speed of stars in the whole spheroid from an averaging of the Doppler effect for the larger system, providing a measure of what is called the velocity dispersion. This is a quite separate measurement, which can be used to reveal the mass of the spheroid in the same way that the motion of galaxies in a cluster reveals the mass of the cluster. Putting everything together shows that more massive black holes live in more massive spheroids. This is not really surprising. The surprise is that the correlation between the two masses is very precise – the central black hole always has a mass of 0.2 per cent of the mass of the spheroid.

This is such a tiny proportion of the total spheroid mass that it also demonstrates clearly that the black hole itself is not responsible for how fast the stars in the spheroid move; all that they 'notice', gravitationally speaking, is their own total mass (i.e. the combined mass of the stars and any remaining clouds of gas and dust between the stars). In essence the spheroid doesn't even know the black hole is there – if you took it away, the galaxy would look and behave exactly the same way.

Although the correlation is most simply expressed in terms of mass, the more significant aspect is that the stars in the spheroid around a more massive black hole move faster. This is an indication that the cloud of baryonic material from which they formed collapsed more within its dark matter halo during the process of galaxy formation. In other words, black holes grew bigger in systems which collapsed more, suggesting that the collapse feeds the black hole as it grows. Black hole masses are determined by the collapse process. It seems very unlikely that supermassive black holes formed first and then galaxies grew around them; they must have formed together, in a process sometimes referred to as co-evolution, from the seeds provided by the original black holes of a few hundred solar masses and the raw materials of the dense clouds of baryons in the knots in the filamentary structure.

The details of how the this symbiotic co-evolution occurred have yet to be unravelled, but it is easy to see in general how the energy pouring out from a growing black hole will first influence the way in which stars form in the surrounding material and then shut off the growth and activity of the black hole at some critical juncture by pushing the surrounding clouds of gas and dust away, simultaneously shutting down the early phase of rapid star formation. This fits observations of starburst galaxies in which winds carrying as much as a thousand solar masses of material a year are seen flowing out from the central regions; such winds, while they last, will trigger star formation in dense interstellar clouds, which they squeeze as they blow upon them. While 0.2 per cent of the available mass gets swallowed by the black hole, about 10 per cent of the baryonic material gets turned into stars.

This relationship between central black hole mass and velocity dispersion holds for a range of black hole masses at least from a few million to a few billion times the mass of the Sun – across a factor of a thousand (three orders of magnitude). It also holds across the Universe from the present day out to at least a redshift of 3.3, when the Universe was only two billion years old. When this relationship was first discovered, it seemed that flat disc galaxies without a central bulge did not have central black holes either, but in 2003 astronomers discovered a black hole with a mass of between 10,000 and 100,000 times the mass of our Sun in the bulge-less disc galaxy NGC 4395. This is still supermassive compared with the Sun, but a flyweight compared with the kind of objects I have been describing so far. But although this galaxy doesn't have a bulge, there is a central concentration of stars with a velocity dispersion that would imply a black hole mass of about 66,000 times that of the Sun. In other words, the velocity dispersion and mass match the relationship found in much larger systems. It may be that all disc and elliptical galaxies harbour central black holes, although irregulars may not.

The relationship also holds for our own Galaxy, the Milky Way, and its near neighbour M31, the Andromeda galaxy. The black hole at the heart of the Milky Way has a mass of only three million solar masses, and there is a small central bulge; the mass of the black hole at the heart of the Andromeda galaxy is 30 million solar masses, and there is a correspondingly larger central bulge. The overall relationship between the Milky Way and the Andromeda galaxy also offers a clue to what happened to galaxies after they had formed along with their central black holes in the early Universe.

The processes I have described so far explain the origin of the smaller elliptical galaxies and the disc galaxies. But the giant ellipticals seem to have formed, as I have already hinted, through mergers of smaller galaxies. At present, the Milky Way and the Andromeda galaxy are moving together at a closing speed of hundreds of km per second. The two galaxies are not destined for a head-on collision, but within at most about 10 billion years they will have merged together to make one giant elliptical. There is some evidence that the Andromeda galaxy has grown to its present size by swallowing a moderately large companion, since it seems to have a double-core, but the anticipated merger between two full-blown disc galaxies will be much more spectacular.

As I have mentioned, stars are so far apart from one another, compared with their own diameters, that even if two galaxies do collide head-on there is very little chance of stellar collisions. The galaxies pass right through each other, with gravity distorting the shapes of the galaxies as it changes the orbits of their stars. Collisions do occur between giant clouds of gas and dust between the stars, and these clouds are also squeezed and distorted by gravitational effects, causing the waves of star formation seen in many starburst galaxies. Gas and dust ejected from each galaxy as they pass through each other will make streams of material within which new globular clusters may form. Then, the galaxies swing round one another and experience another interaction. The process continues, with the cores of the galaxies getting closer on each swing, until the two galaxies merge into a single system in which there is no obvious disc but a whole mass of stars within which there are streams moving at various orientations, some of them carrying a memory of the discs that used to be. The final merger of the two central black holes releases a blast of energy which triggers a final phase of starburst activity before the new giant elliptical settles down into a quiet life. The penultimate stage of such a merger can actually be seen in the galaxy NGC 6240, where there are two black holes a kiloparsec or so apart, moving together on a collision course at the heart of the galaxy.

It used to be thought that in the case of the Milky Way and the Andromeda galaxy the timescale for all this ran from about five billion years to about 10 billion years from now, after the Sun will have ended its life as a bright star. But in 2007 a team at the Harvard-Smithsonian Center for Astrophysics reported calculations which suggest that the distortion of the Milky Way could begin in only about two billion years, when there could conceivably be intelligent life in our Solar System to watch the events. Any such watchers would have to be extremely patient, though, since even on the basis of the revised timescale the merger will then take a further three billion years to complete. By that time, the ageing Sun would have been displaced into an orbit about 30 kiloparsecs from the centre of the merged system, roughly four times more than its present distance from the centre of the Milky Way. Although the jury is still out on whether this revised timescale should be accepted, the end result will be much the same, whenever it happens.

Close encounters can also cause galaxies to shrink. In rich clusters of galaxies the individual members (the 'bees' in the 'swarm') are moving so fast under the influence of gravity that they cannot merge, but sweep past each other in glancing encounters that strip dust and gas, and even stars, from each other, sending the material flowing out into intergalactic space, where it forms a hot fog which can be detected at X-ray wavelengths. The largest elliptical galaxies sit at the centres of such clusters, like a spider sitting at the heart of its web, devouring anything that comes near them and growing fatter as they do so.

About one in a hundred of the galaxies seen at low redshifts are actively involved in the late stages of mergers, but these processes take so little time, compared with the age of the Universe, that the statistics imply that about half of all the galaxies visible nearby are the result of mergers between similarly sized galaxies in the past seven or eight billion years. Disc galaxies like the Milky Way seem themselves to have been built up from smaller sub-units, starting out with the spheroid and adding bits and pieces as time passed. I have already mentioned the star streams that are interpreted as remnants of lesser objects captured by our Galaxy; another line of evidence that supports this idea, probing further into the past, comes from globular clusters, whose ages can be inferred with good accuracy by studying their composition using spectroscopy.

The first stars contained very little in the way of elements heavier than hydrogen and helium, while younger stars are enriched with elements manufactured inside previous generations of stars in a well-understood way. Each globular cluster is made up of stars with the same age, confirming that they formed together from a single cloud of gas and dust. But the clusters have different ages from each other, showing that they formed at different times. The oldest are a little over 13 billion years old, nicely matching our understanding of when the first galaxies formed. The spread of ages of the globular clusters supports the idea that the part of our Galaxy outside the original spheroidal bulge formed from hundreds of thousands of smaller gas clouds each with up to a million solar masses of material. Whenever a new gas cloud collided with the growing galaxy, it would send a shock wave rippling through the cloud and trigger a burst of star formation in its core, forming a new globular cluster. The bulk of the material from the cloud would be tugged by gravity and slowed by friction to become part of the growing disc of material around the spheroidal bulge. Some globular clusters would survive until the present day; others would get ripped apart by tidal forces if their orbits took them too deep in towards the centre of the Galaxy. But the computer simulations show that this whole settling down process only works on the timescale available, if at all, if there is dark matter contributing to the overall gravitational field – several times more dark matter than there is baryonic matter. Without dark matter, disc galaxies could not grow at all, and without dark matter there would not have been any spheroidal seeds for them to grow on in the first place.

Within this self-consistent framework, small irregular galaxies are simply seen as bits and pieces left over from the early days of the Universe. Although it is difficult to see smaller galaxies far away, it is possible to make allowances for this in interpreting the statistics; when this biasing is allowed for, the observations tell us that there were many more small galaxies when the Universe was young than we see around us today. This is exactly what we would expect if many of the small galaxies have either grown larger through mergers or been swallowed up by larger galaxies. At the other extreme, more than half of the mass of baryonic material in the Universe today has already been converted into giant elliptical galaxies, the largest of which contain several trillion (10^{12}) times as much mass as our Sun – equivalent to about ten galaxies like the Milky Way put together. These are seen as far back as a redshift of 1.5, but spectroscopic studies reveal that many of them were fairly old by then, and that the components from which they formed must have merged at redshifts of 4 or more. Nevertheless, although the great era of galaxy mergers may have happened more than 10 billion years ago, probably the most important point is that these processes are still going on today. Galaxies are still involved in interactions and mergers, and clusters are still forming into superclusters. In this sense, the Universe of galaxies is still young; it has yet to mature. But what will be the ultimate fate of galaxies?

Chapter 8
The fate of galaxies

The fate of galaxies depends upon the fate of the Universe. There are three basic scenarios to be considered, and although theorists have come up with many variations on the basic themes these subtleties do not significantly alter the three possibilities for the fate of galaxies. The first possibility is that the Universe will keep expanding in more or less the same way as today, with a steady acceleration. The statistics of observations available today favour this possibility, but not decisively enough to rule out the two other options. The second possibility is that the acceleration of the expansion rate will itself accelerate. The third is that the acceleration will switch over into a deceleration at some point in the not too distant future and the Universe will eventually collapse into a Big Crunch that is the time-reversed version of the Big Bang.

All these scenarios are speculative, and when we look at the timescales involved there is no point in talking in anything but round numbers, so we start with the present age of the Universe rounded off to 10 billion (10^{10}) years as a benchmark. Also, we know so little about the nature of dark matter that it is difficult even to speculate about what might happen to it in the distant future, so I shall concentrate on the fate of baryons, the familiar particles that we ourselves are made of.

If the expansion of the Universe continues for long enough, then eventually all of the available gas and dust will be used up, and star formation will cease. From studies of the history of star formation, revealed by observations of different populations of stars in nearby galaxies, and the rate at which stars are being formed today in our own Galaxy, astronomers infer that this will happen in about a trillion (10^{12}) years from now – when the Universe is a hundred times older than it is today. Individual galaxies will become redder and dimmer as their stars fade and cool, and galaxy clusters will be carried ever farther apart, making it impossible for any astronomers around at the time to look out into the Universe and see anything beyond their own cluster. As the stars within each galaxy die, they will settle into one of three states. Stars with masses similar to or less than that of the Sun will simply fade away into cinders called white dwarfs, lumps of star-stuff containing about as much mass as our Sun in a sphere that is as big as the Earth. Stars which end their days with slightly more mass than this will shrink even further, forming compact balls with the mass of the Sun squeezed into the volume of Mount Everest; such neutron stars contain material at the density of an atomic nucleus. If there is even more mass left over when a star dies, or if a neutron star accretes enough mass from its surroundings, it will collapse all the way into a black hole.

Galaxies also shrink, on these very long timescales. This is partly because they lose energy through gravitational radiation, which has a trivial effect on any human timescale but adds up over trillions of years. They also shrink because of encounters between stars in which one star gains energy and is ejected into intergalactic space, while the other loses energy and falls into a tighter orbit around the galactic centre. In the same way, clusters of galaxies will shrink, and eventually both individual galaxies and clusters of galaxies will fall into supermassive black holes built up by this process.

You could regard this as the end of the story, since nothing recognizable as a galaxy would still exist at that time. But there will still be black holes and baryons around, in the form of ejected stars and traces of gas. If there is enough time, then according to particle physics theory even these ultimate constituents of the Universe will disappear. To indicate the timescales involved, for the moment I shall ignore the cosmological constant, and look at the old picture of a universe expanding steadily but more slowly as time passes, giving us infinite time to consider.

Theory tells us that the same processes which made matter out of energy in the Big Bang would eventually turn matter into energy as the Universe ages. 'Eventually' is the key word. Atoms are made of three kinds of particle: electrons, protons, and neutrons. Electrons are truly fundamental, stable particles; but neutrons, left on their own outside an atom, will decay into a proton and an electron in a few minutes. Protons seem to be stable on timescales comparable to the present age of the Universe, but theory says that even protons decay eventually, each one turning into a positron (the antimatter equivalent to an electron) and energetic gamma rays. Something similar happens to neutrons in white dwarfs and neutron stars, in this case with each decay producing both an electron and a positron to keep the overall balance of electric charge. The equations that describe how matter was produced in the Big Bang suggest that in any lump of ordinary matter half of the protons will decay in about 10^{32} years. Turning this around, in a lump of matter containing 10^{32} protons, one will decay every year or so. This is about the number of protons in 500 tonnes of anything – including water, butter, or steel.

This is a mind-bogglingly long time. Even 10^{30} is 10 billion cubed – a thousand billion billion billion – and 10^{32} years is a hundred times longer than 10^{30} years. But by about 10^{33} years from now, if the steady expansion carries on that long, virtually all the baryons not already swallowed up by black holes will have been processed in this way into electrons, positrons, and energy.

Whenever an electron and a positron meet, they annihilate each other in a puff of gamma rays. So all of the leftover star-stuff eventually ends up as radiation.

What of the black holes? Curiously, they suffer the same fate. There is a profound connection between the description of a black hole in terms of the general theory of relativity and thermodynamics and quantum theory. The key to this is the principle that lies at the heart of quantum physics, known as quantum uncertainty. This tells us that there are certain pairs of properties in the quantum world which match up in such a way that it is impossible for both members of the pair to have a precisely defined value at the same time. This is nothing to do with the imperfections of our measuring equipment, but is a feature of the way the Universe works. One pair of such variables is energy and time. In the context of the fate of black holes what matters is that the energy/time uncertainty tells us that there is no such thing as truly 'empty' space. If you imagine a tiny little volume of empty space, you might think it contained zero energy. But quantum uncertainty tells us that it *might* contain a certain amount of energy, E, provided it only does so for less than a certain time t. The bigger E is, the smaller t must be. So a little bubble of energy can pop into existence, then vanish, without ever being detected. Since energy can be equated with mass, this means that a pair of particles, such as an electron and a positron, could pop into existence out of nothing at all, provided they promptly disappear again.

Suppose this happened right at the edge of a black hole. Even in the tiny time available, one member of the pair could be captured by the black hole, while the other one escapes. But the Universe has not gained something for nothing; some of the mass of the black hole has been used up in the process, and the hole shrinks by a minuscule amount. The resulting avalanche of particles away from the surface of a black hole gives it a well-defined temperature, which is where thermodynamics comes into the story. The way the effect works, small black holes are hotter, and will evaporate away to nothing at all, exploding in a burst of radiation at the point where the mass inside the hole is no longer enough to close itself off from the rest of the Universe. A black hole with the mass of the Sun would take 10^{66} years for this to occur, even if it never swallowed any outside matter along the way. A black hole with the mass of a galaxy will evaporate in 10^{99} years, and even a hole containing the mass of a supercluster of galaxies – the biggest ever likely to form – will be gone in 10^{117} years. That really is as far as we can push our speculations and still pretend we are talking about the fate of galaxies.

But what if there isn't time for all this to happen? If the cosmological constant really is constant, and the expansion of the Universe accelerates at a constant rate, everything beyond the Local Group of galaxies, of which our Milky Way Galaxy is a member, will be carried out of sight within a couple of hundred billion years. Space outside our local bubble will be expanding faster than light, and no signal from outside will ever be able to reach any observers in the Milky Way, or whatever the Milky Way has become. There will be, in effect, a shrinking cosmic horizon defining the limit of observations. The processes I have just described will still go on, both outside this bubble and within it, but for all practical purposes within about 10 times the present age of the Universe there will be nothing to see outside the fading island of stars represented by whatever kind of merged supergalaxy has formed from the components of the Local Group. This is today's 'best buy' in terms of astronomical prognostications. There are, though, more dramatic possibilities. What if the cosmological 'constant' isn't really constant at all?

The supernova studies set limits on how much the cosmological constant could have changed as the Universe has evolved, but they are not yet good enough to prove that it really has been constant ever since the Big Bang. Perhaps it should really be called the cosmological parameter, to allow for the possibility that it changes as time passes. This has encouraged some theorists to speculate about how a changing value for the dark energy density of the Universe would affect the stretching of space and the fate of galaxies. The first possibility, that the rate at which the expansion of the Universe is accelerating may itself be accelerating, completely changes our view of our place in the Universe, since it suggests that, far from living in an early stage of a universe destined for a long life, we may already be a third of the way from the Big Bang to the end of everything material. Even more intriguingly, this idea suggests that if intelligence survives in the universe, observers will be able to watch this ultimate destruction almost to the end. (I use the lower case for 'universe' here to highlight that these are speculations about a possible universe, not certainties that apply to our Universe. My personal view is that these are fantasies, but entertaining ones!)

This scenario is sometimes referred to as the Big Rip, for reasons that will soon become obvious. It starts from the assumption that the expansion of the universe is responsible for creating dark energy, while, as I have explained, dark energy makes the universe expand faster. More expansion implies more dark energy, which implies a faster expansion, which implies more dark energy, and so on. All this is consistent with the known laws of physics, but is not required by those laws. If the cosmological parameter stays as small as it is today, objects like the Sun and stars, and galaxies, have no trouble resisting the cosmic expansion for hundreds of billions of years, because their gravity overwhelms the effects of dark energy. But in the runaway Big Rip scenario there soon comes a time when the dark energy, acting as an ever more powerful kind of antigravity, overwhelms gravity, and even what we think of as solid objects get torn apart by the expansion. This is an example of exponential growth, and, even in the most extreme version of the Big Rip allowed by the observations, although the end will occur in a little over 20 billion years time, nothing particularly odd will happen to objects like galaxies until the last billion years or so.

At that time, dark energy will overpower the gravitational forces holding the Local Group of galaxies together; this happens about 20 billion years from now, 10 times sooner than this will happen if the cosmological constant really is constant. By then, the large elliptical galaxy formed by the merger between the Milky Way and the Andromeda galaxy will still exist in a recognizable form, and although the Sun will have been dead for well over 10 billion years, there could very well be intelligent beings living on other Earth-like planets orbiting Sun-like stars, able to watch what happens as the size of the cosmological parameter continues to increase; the cosmological 'horizon' at that time would still be at a distance of about 70 Megaparsecs.

From this point on, it makes sense to measure the passage of events not in terms of the time that has elapsed since the Big Bang, but in terms of the time left before the Big Rip. Some 60 million years before the end, the galaxy – and all galaxies – would begin to evaporate as the dark energy became strong enough to overcome the gravitational attraction between stars, but it would still be possible for a planetary system like the Solar System to wander through space intact. Just three months from the Big Rip, the gravitational bonds holding planets to their parent stars would be loosened, and even any civilization that had the technology for observers to survive that catastrophe would reach its end when their planet was ripped apart by the cosmic expansion, about half an hour before the end of matter. In the last fraction of a second, atoms and particles would be ripped into nothingness, leaving a flat and empty spacetime. Some extreme versions of this idea suggest that a new universe might be born out of this void, and that our own Universe may have come from such a void. But as far as galaxies are concerned, we can say that the end is due, if this scenario is correct, in about 20 billion years' time and about 60 million years before the Big Rip.

Suppose, though, that the cosmological parameter fades away as time passes. It could fade away to zero, which would give us back the image of an ever expanding universe, with decaying matter and evaporating black holes, with which I started this overview. But why stop there? The equations allow for the possibility that the parameter could become negative. That brings doomsday even closer, perhaps as close to us in the future as the Big Bang is to us in the past. But this would be a different kind of doomsday – not a Big Rip, but a Big Crunch, equivalent to the Big Bang run in reverse.

Once again, I'll use the most extreme version of the scenario consistent with our observations of the real Universe and the known laws of physics. Just as a positive amount of dark energy acts like antigravity and makes the universe expand faster, a negative amount of dark energy acts like gravity and pulls the universe together, possibly reversing the cosmic expansion. Observations made so far, combined with theoretical considerations, suggest a range of possibilities for this kind of decline in the value of the cosmic parameter, implying that the Big Crunch could happen in as little as 12 billion years from now or as far into the future as 40 billion years from now. As before, the events are best described in terms of the time left before the end, which can also be expressed in terms of the shrinking size of the observable part of the universe. Since everything shrinks in the same way, even outside our horizon, exactly the same processes will be going on everywhere at the same time. But this time, intelligent observers would not be around to witness the death throes.

When the universal expansion halts and then reverses, it affects everywhere in the universe at the same time, because it is space itself that is affected by the changing value of the cosmological parameter. But because of the finite time it takes light to travel through space, any observers around just after the reversal, wherever they may be in the universe, would not see a universe dominated by blueshifted galaxies. Light from nearby galaxies would be blueshifted, but the light from distant galaxies, which had spent most of its journey travelling through expanding space, would still be redshifted. A long-lived civilization would be able to keep records showing the spread of a 'blueshift horizon' outwards at the speed of light, until eventually blueshifts do indeed dominate.

As far as galaxies are concerned, the collapse of the universe will have little effect for billions of years. The processes of star formation and galactic mergers that I have described will carry on just as before, with clusters of galaxies falling towards each other and eventually merging, and galactic mergers becoming ever more common, but still without any major problems for life forms like us living on planets like the Earth. The threat to life actually comes from one of the feeblest features of our Universe today: the background radiation left over from the Big Bang.

This cosmic microwave background radiation is left over from the fireball in which the Universe was born. Between 300,000 and 400,000 years after the Big Bang, at the time of decoupling, it was as hot as the surface of a star is today, and it has cooled all the way down to a temperature of 2.7 K (−272.3 °C) as it has stretched to fill the space available. But when the space available shrinks, the radiation will be blueshifted and compressed, heating up in the exact reverse of the process that cooled it. At the time when clusters have started to merge and all galaxies are beginning to be involved in mergers, the universe will be about one hundredth of its present size and the temperature of the sky will be about 100 K, still not enough to be alarming. But within a few million years, the temperature of the background radiation will exceed the melting point of ice, 273 K, and there would be no more snow or ice anywhere in the universe. Life might still be possible, but as the temperature continues to increase it moves up past the boiling point of water, 373 K, and soon the whole sky begins to glow brighter and brighter as time passes.

Life becomes impossible, and galaxies are disrupted into a mess of stars, a couple of billion years before the Big Crunch. A little less than a million years away from the end all baryonic matter that is not safe inside stars 'decombines' into its electrically charged components. Now, matter and radiation are recoupled into an intimate embrace. This is exactly the reverse of the decoupling that occurred after the Big Bang, and it happens at exactly the same time, 300,000 to 400,000 years away from the end, as decoupling occurred away from the beginning. The difference is that stars, or at least their cores, can still survive within this fireball, until the universe reaches one millionth of its present size and the temperature exceeds 10 million degrees, comparable to the temperatures inside stars. Then, even the stellar cores dissolve into the fireball. Eventually, everything disappears into a singularity, like the singularity at the heart of a black hole, or the one from which the Universe was born.

Which leads to the intriguing speculation that our Universe may have been born in *exactly* this way, out of the collapse of a previous universe, or a previous phase of our own Universe, which may follow a repeating cycle of expansion, collapse, and bounce. None of that, though, is relevant to the fate of the galaxies we see in our Universe. In the Big Crunch scenario, galaxies as we know them would be disrupted beyond recognition by about a billion years from the end, perhaps only 11 billion years from now.

But both the Big Rip and Big Crunch scenarios are speculations, offered here primarily to show the limits of what could happen. As far as we can tell, it is not possible for the Universe to recollapse in less than about 12 billion years, and nor is the Big Rip going to tear galaxies apart within the next 20 billion years. Thirty years ago, there was almost exactly the same uncertainty, between 12 billion years and 20 billion years, in astronomers' estimates of the time that has elapsed since the Big Bang. This has now been pinned down to 13.7 billion years. That's progress.

Maybe we can hope for similar progress over the next 30 years concerning our understanding of the fate of the Universe.

The present best buy prognostication for the fate of galaxies, however, is that the cosmological constant really is constant, and that although as a result of the gradual acceleration in the expansion rate of the Universe a Slow Rip may eventually happen, it will be so far in the future that it is scarcely worth bothering about. On that picture, galaxies are safe for a few hundred billion years, more than 10 times the present age of the Universe, and there will be plenty of time for other intelligent observers to work out exactly how it will all end.

Glossary

accretion disc: A disc of material orbiting around a star, black hole, or other object, from which matter spirals inward to fall onto the central object.

black hole: Any object with a gravitational pull so powerful that the escape velocity exceeds the speed of light. Supermassive black holes are the seeds of galaxy formation.

Cepheid: A kind of variable star whose properties make it useful in calculating distances across the Milky Way and to nearby galaxies.

Cold Dark Matter: The dominant material component of the Universe, present in about the ratio 6:1 compared with everyday matter. The presence of CDM is revealed by its gravitational influence, but nobody knows exactly what it is.

cosmological constant: A number which indicates the amount of dark energy in the Universe.

dark energy: Invisible form of energy, also known as the lambda field, thought to fill the entire Universe and act as a kind of antigravity, increasing the rate at which the Universe expands.

disc galaxy: A system of hundreds of billions of stars, most of them in a flattened disc, where there may be a spiral structure. Our Milky Way is a disc galaxy.

Doppler effect: Shift in the lines of a spectrum (for example, of a star) towards the red end of the spectrum if it is moving away, and towards the blue if it is moving towards us.

elliptical galaxy: A large system of stars with no obvious internal structure, with an overall shape like that of an American football.

escape velocity: The minimum speed needed for an object to escape from the gravitational clutches of another object. The escape velocity from the surface of the Earth is 11.2 km per second.

extinction: The dimming of light from distant stars by dusty material along the line of sight.

Galaxy (capital 'G'): The galaxy in which we live; also known as the Milky Way.

galaxy (small 'g'): Any one of the hundreds of billions of islands of stars in the Universe.

globular cluster: A spherical ball of stars found in the outer regions of a galaxy like the Milky Way. A single globular cluster may contain millions of individual stars.

Hubble Constant: A number which specifies how fast the Universe is expanding today. The rate of expansion changes as time passes.

Lambda (Λ) field: *See* dark energy.

Milky Way: A band of light across the night sky made up of vast numbers of stars too distant to be seen individually with the naked eye. *Also see* Galaxy.

nova: The sudden brightening of a star which makes it look like a 'new' object in the sky.

nuclear fusion: The process of fusing light nuclei (in particular, those of hydrogen) into heavier nuclei (in particular, those of helium). This releases energy and keeps stars like the Sun shining.

parallax: The apparent movement of an object across the sky when seen from different positions.

principle of terrestrial mediocrity: The idea that we do not occupy a special place in the Universe and that our surroundings are typical of those of a star in a disc galaxy.

spectroscopy: Technique of analysing the light from stars or galaxies by spreading it out into a spectrum.

spiral galaxy: *See* disc galaxy.

supernova: Extreme brightening of certain kinds of star at the end of their lives, when the single star can shine for a short time as brightly as a whole galaxy of stars like the Sun.

Universe (capital 'U'): The totality of everything we can see or be influenced by – the 'real world'.

universe (small 'u'): Term used to refer to a theoretical model, based on calculations and/or observations of what the world we inhabit might be like.

Further reading

Richard Berendzen, Richard Hart, and Daniel Seeley, *Man Discovers the Galaxies* (Columbia UP, 1984).
Peter Coles, *Cosmology: A Very Short Introduction* (OUP, 2001).
Arthur Eddington, *The Expanding Universe* (CUP, 1933)
John Gribbin, *Space* (BBC Worldwide, 2001).
John Gribbin, *Science: A History* (Allen Lane, 2002).
Alan Guth, *The Inflationary Universe* (Cape, 1996).
K. Haramundanis ed. *Cecilia Pagne-Gapschkin: An Autobiography and Other Recollections* (Cup, 1984)
Michael Hoskin, 'The Great Debate', *Journal for the History of Astronomy*, 7 (1976), 169–82.
http://antwrp.gxfc.nasa.gov/apod/ (for the observations in Hawaii, Chapter 3).
Edwin Hubble, *The Realm of the Nebulae*, Dover, 1958 (repr. of 1936 edn).
Malcolm Longair, *Our Evolving Universe* (CUP, 1996).
Denis Overbye, *Lonely Hearts of the Cosmos* (HarperCollins, 1991).
Martin Rees, *Before the Beginning* (Simon & Schuster, 1997).
Michael Rowan-Robinson, *The Cosmological Distance Ladder* (Freeman, 1985).
Thomas Wright, *An Original Theory or New Hypothesis of the Universe* (Chapelle, 1750; facsimile edn, ed. Michael Hoskin, Macdonald, 1971).

Index

A

B

C

D

E

F

G

H

I

K

L

M

N

O

P

Q

R

S

T

V

W

激发个人成长

多年以来，千千万万有经验的读者，都会定期查看熊猫君家的最新书目，挑选满足自己成长需求的新书。

读客图书以“激发个人成长”为使命，在以下三个方面为您精选优质图书：

1. 精神成长

熊猫君家精彩绝伦的小说文库和人文类图书，帮助你成为永远充满梦想、勇气和爱的人！

2. 知识结构成长

熊猫君家的历史类、社科类图书，帮助你了解从宇宙诞生、文明演变直至今日世界之形成的方方面面。

3. 工作技能成长

熊猫君家的经管类、家教类图书，指引你更好地工作、更有效率地生活，减少人生中的烦恼。

每一本读客图书都轻松好读，精彩绝伦，充满无穷阅读乐趣！

认准读客熊猫

读客所有图书，在书脊、腰封、封底和前后勒口都有“**读客熊猫**”标志。

两步帮你快速找到读客图书

1. 找读客熊猫

2. 找黑白格子

马上扫二维码，关注“**熊猫君**”

和千万读者一起成长吧！